SUSTAINABLE USE OF GENETIC RESOURCES UNDER THE CONVENTION ON BIOLOGICAL DIVERSITY

Exploring Access and Benefit Sharing Issues

To Lydia, with love

Sustainable Use of Genetic Resources under the Convention on Biological Diversity

Exploring Access and Benefit Sharing Issues

W. LESSER

Department of Agricultural, Resource, and Managerial Economics
Cornell University
USA
and
International Academy of the Environment
Switzerland

CAB INTERNATIONAL

CAB INTERNATIONAL
Wallingford
Oxon OX10 8DE
UK

CAB INTERNATIONAL
198 Madison Avenue
New York, NY 10016-4314
USA

Tel: +44 (0)1491 832111
Fax: +44 (0)1491 833508
E-mail: cabi@cabi.org

Tel: +1 212 726 6490
Fax: +1 212 686 7993
E-mail: cabi-nao@cabi.org

A catalogue record for this book is available from the British Library,
London, UK.

Library of Congress Cataloging-in-Publication Data
Lesser, William.
    Sustainable use of genetic resources under the Convention on
Biological Diversity: exploring access and benefit sharing issues /
W. Lesser.
       p. cm.
    Includes bibliographical references and index.
    ISBN 0-85199-197-1 (alk. paper)
    1. Biological diversity conservation—Law and legislation.
2. Genetic engineering—Law and legislation. 3. Sustainable
development—Law and legislation. I. Convention on Biological
Diversity (1992) II. Title.
K3488.L47   1998
333.95'11–dc21   98–23972
ISBN 0 85199 1971   CIP

Typeset in Plantin by Advance Typesetting Ltd, Oxford
Printed and bound in the UK by Biddles Ltd, Guildford and King's Lynn.

Contents

Preface

Three years have passed since the Convention on Biological Diversity came into force in record time, and slightly longer since the Rio Earth Summit created such enthusiasm and expectations. The judgement as to whether the Convention has met many of those expectations is a personal one, based in a large measure on one's realistic anticipations and familiarity with the functioning of international conventions. Clearly, the Convention is a complex one, not only in the language, where it certainly is, but also in the attempted balance between conservation and sustainable use, and between the public and private sectors. As clearly, the drafters were absolutely correct that all those elements are required to achieve the stated objectives.

Now, the parties are striving to implement those objectives, and I have the temerity to attempt, with this volume, to assist that process, even in a small way. Not being completely unrealistic, I have limited my scope to matters of access and use on mutually agreed terms. Attempting to cleave off those components is somewhat hollow, for conservation is essentially bypassed, or worse, relegated to the general expectation that, human nature being what it is, we conserve what has evident value. Yet a fuller integration is beyond my powers for even my limited scope leaves me with seven key articles to evaluate, plus fragments of the Nairobi Final Act. What I do attempt to contribute is a pragmatic economic perspective to issues which, at their core, will often rely on economic incentives to achieve the objectives.

Suppliers of genetic resources, like suppliers of technologies, understandably wish to be paid equitably for their contributions. Why should that be otherwise? The Convention will not change those basic expectations, more so in the current period of 'privatization' worldwide. Yet, for many of these materials, no proper markets exist. We often do not even know what exists, let alone how to value and transfer it. That is where applied economics comes in, not as an absolute guideline, but as a perspective on the past and, through that, a guideline to the future, however limited and partial it may be. In short, much of the Convention on biological diversity relates to supply and demand, the core of economics, and here I attempt to apply economic thinking to the resolution of some of the sticking points of the discussions. In my judgement, such a pragmatic perspective is often missing from the cloud of words engulfing the Convention process.

Often my perspective is an individual one. That is to say, I seek partial answers where the full scope of the matter seems to exceed a resolution. That has led me, for example, to suggest that the quest for rights to indigenous/local knowledge be separated from the rights to the underlying genetic resources themselves. In terms of economics that presents no conceptual problems, but it may for those whose full objective is the restoration of rights to traditional lands. As worthy a goal as that may be, it is not fundamentally an economics issue and hence is excluded from consideration here. Some (hopefully) will like my approach, others not. Some will applaud the implicit and, in places, explicitly stated need for greater private sector involvement, while others deplore it. The most I can hope for is to provide some insights into cause and effect; if one seeks a particular market-oriented goal, economic theory and experience suggest one proceeds in the identified ways. That is, herein I do not propose that countries and/or individuals should seek to commercialize sustainably their genetic resources. That is their decision. Rather, once that decision has tentatively been made, I explore means of implementation and their consequences.

In places, I have indulged myself with some repetition. This occurs, for example, with the descriptions of intellectual property rights in Chapters 2, 7 and 9. Hopefully, the reader will be sufficiently patient with me to recognize that different aspects are explored in each chapter. When duplication does occur, it is provided in an attempt to make the chapters self-contained. A few readers may read this work cover to cover, but that is not something I would rely on.

Will my proposals and recommendations work in the case of the Biodiversity Convention? Who really knows, but it is my fond hope that some are given the opportunity to be tested in practice. We – the public, diplomats, scientists, and social scientists, both traditional and formally trained – are all groping with the proper management of biodiversity. Yet we have learned from the diversity of the world that multiple approaches are essential. May mine be considered among them.

W. Lesser
Ithaca, New York
April 1997

Acknowledgements

To borrow a phrase from a colleague, a turtle doesn't get to the top of a fence post without help, and so it is with the completion of this volume. My secretary, Jonell Blakeley, worked tirelessly and skillfully in the massaging of numerous drafts, all at a time in the course of Cornell University budget constraints when her work-load effectively doubled. Ellen McCallie, a graduate student in SCAS, out of the goodness of her heart, provided me with editorial and substantive suggestions.

Numerous others assisted in aspects of this work. Manuel Ruiz Muller provided successive drafts of the Andean Pact access legislation, plus commentary; my graduate students Juan Lucas Restrepo and Alejandro Lalor assisted with translation and interpretation. Octavio Espinosa responded graciously to an eleventh hour appeal for documents on the amended copyright treaty. Concurrently, I completed a small project for IPGRI, parts of which found their way into this book. During that process I enjoyed, and profited from, working with Marcio de Miranda Santos and IPGRI staff, including George Ayad, Jan Engels, Ruth Raymond, and Masa Iwanaga. Joseph Straus kindly provided some current documents not previously known to me.

I drew especially heavily from several excellent reports and would be remiss in not acknowledging them. They included the FAO 'Draft Report on the State of the World's Plant Genetic Resources', and the UNEP *Global Biodiversity Assessment*, for which V.H. Heywood served as Executive Editor, but many individuals (including me, in a very small way) contributed. It represents a truly amazing effort, and contribution.

None of this should be interpreted as saying any lingering errors are anyone's responsibility but my own.

Abbreviations and Acronyms

ABSP   Agricultural Biotechnology Support Program
ACTS   African Center for Technology Studies
CBD   Convention on Biological Diversity
CGIAR   Consultative Group for International Agricultural Research
COP   Conference of the Parties of the Convention on Biological Diversity
EPC   European Patent Convention
EPO   European Patent Office
FAO   Food and Agricultural Organization (of the United Nations)
GATT   General Agreement on Tariffs and Trade
GPA   Global Plan of Action
IARC   International Agricultural Research Centers
INBio   National Biodiversity Institute (Costa Rica)
IPGRI   International Plant Genetic Resources Institute
IPR   Intellectual Property Rights
IRRI   International Rice Research Institute
ISAAA   International Service for the Acquisition of Agri-biotech Applications
IUCN   World Conservation Union
IUPGAR   International Undertaking on Plant Genetics Resources
MFN   Most Favoured Nation
MTA   Material Transfer Agreements
MUSE   MUltilateral System for Exchange
NGO   Non-Governmental Organization
NPV   Net Present Value
OECD   Organization for Economic Cooperation and Development
PBR   Plant Breeders' Rights
PIC   Prior Informed Consent
PPC   Production Possibility Curve

SBSTTA   Subsidiary Body for Scientific, Technical and Technological
          Advice (of the COP)
SINGER   System-wide Information Network on Genetic Resources
TEV   Total Economic Value
TRIPs   Trade Related Aspects of Intellectual Property Rights
UNCTAD   United Nations Conference on Trade and Development
UPOV   International Union for the Protection of New Varieties of
          Plants
VMP   Value of Marginal Product
WIPO   World Intellectual Property Organization
WTA   willingness to accept (compensation to give up what presently
          exists)
WTO   World Trade Organization
WTP   willingness to pay

Chapter 1

Introduction

Most people, specialists and non-specialists alike, struggle with meaningful definitions of *genetic resources* and *biodiversity*, as well as their interrelationship. Scientists indeed have experienced great difficulty estimating the number of types of living organisms to within even an order of magnitude. Generally accepted estimates range from 3.6 million to 111.6 million species, with a 'working' figure of 13.6 million (Hawksworth and Kalin-Arroyo, 1995; Table 3.1–2), of which about 1.4 million have been described (Wilson, 1992). What is broadly recognized is that the rate of species extinction is rising, reaching alarming and, by some interpretations, non-sustainable levels.

Estimates of species loss range upwards to 10% for the coming quarter century, or about 1000–10,000 times the 'background' extinction rate preceding major human impacts (literature review in US National Research Council, 1992, Chap. 1; Barbault and Sastrapradja, 1995, Chap. 4.4). Wilson places the rate at 0.5% annually for the next 30 years (quoted in Evenson, 1994). Firm numbers are, of course, impossible while the exact tabulation of species remains so elusive. Particularly troubling for many is the potential loss of: (i) visible species like giant pandas and roniceri, (ii) habitat destruction, especially in species-rich areas like rainforests and centres of origin of agricultural crops, and (iii) crop species, including wild relatives. Numerous expeditions report an inability to relocate species first collected even a few decades ago, while teosinte, the wild maize relative, is maintained at the margins of traditional farmers' fields, vulnerable to changing economic conditions (Wilkes, 1993, unpublished manuscript). Indeed, the use of landraces, traditional cultivars, has been in precipitous decline since the 1960s; where Indian farmers may have once planted 30,000 rice varieties, they are projected soon to plant just ten varieties on 75% of their acreage (Jain, 1982).

This unchecked decline of biodiversity, or, more properly, the loss of diversity-supporting habitat, has many possible explanations. Some lay the

Box 1.1. Common causes of biodiversity decline.

Population growth, urbanization, habitat and clearing for agriculture (Myers, 1988)

Poverty (Myers, 1988)

Household use of wood for cooking

Climate change (Myers, 1988)

Consumerism, leading to logging and conversion of forests to ranches (Myers, 1988)

Habitat destruction, changes in quality and fragmentation (Barbault and Sastrapradja, 1995)

Persecution and exploitation of population (Barbault and Sastrapradja, 1995)

Competition with introduced species (Barbault and Sastrapradja, 1995)

Gross inequity in access to resources; conflicts between native peoples and migrants

World wars and displacement of human populations (McNeely et al., 1995)

Large engineering works (McNeely et al., 1995)

High input agriculture including rapid and widespread adoption of improved cultivars (McNeely et al., 1995)

Large-scale industrial production (McNeely et al., 1995)

Dietary change, especially toward commodity crops and meat (McNeely et al., 1995)

blame on multinational timber companies, while others implicate fuel-wood collectors. Population growth is the paramount cause to one group, yet some imply the problems are traceable to the widespread decline of 'traditional' agricultural practices. Overall, loss is directly associated with degradation and habitat destruction, but what lies behind these? (Box 1.1). In truth, the causes are multiple and interrelated; as such, they are difficult to identify and rectify. In fact, the only certain factor is the need to ameliorate the situation through focused multinational efforts.

Creating that focus has been the objective of multiple conventions over the past decade or so, most held under the auspices of the United Nations. Examples include CITES (Convention on International Trade in Endangered Species of Wild Flora and Fauna), the Montreal Protocol (controlling chlorofluorocarbons (CFCs)) and Desertification (Box 1.2). Each convention is somewhat specific in its objective, taking a relevant approach. CITES operates on the basis of lists of endangered species for which the signatory nations have pledged to ban trade in parts. For example, no longer can rhinoceros-horn daggers be imported into the signatory countries. Under the effective

> **Box 1.2.** Selected recent multinational environmental conventions.
>
> Ramsar Convention (1971): wetland habitats
>
> CITES (1973): trade in endangered species
>
> Montreal Protocol (1987): protect the ozone layer
>
> Framework Convention on Climate Change (1992): CO_2 emissions
>
> Basel Convention (1989): transboundary movements of hazardous wastes
>
> Convention on Biological Diversity (1992): biodiversity preservation and use
>
> Convention to Combat Desertification (1994): stem the expansion of deserts

approach of the Montreal Protocol, a future ban on the production and marketing of CFCs was agreed to, along with more lenient conditions for the developing country signatories.

Each convention protects biodiversity in some aspect, by species, by ecosystem type, or by region. Yet, lacking was a specific approach for the preservation of biodiversity globally in its multiple components. This came through the Convention on Biological Diversity (CBD) (also referred to here by its common, if non-formal, name of the 'Biodiversity Convention'), adopted on 22 May 1992 and opened for signatures at the United Nations Conference on Environment and Development (UNCED; also known as the Earth or Rio Summit) later that year. Approval proceeded extremely rapidly so that the Convention entered into force in December of 1993, 90 days following adoption by the requisite 30 countries. 'Entry into force' means became legally binding; nationally, conventions are treated similarly to treaties which must be approved by each member government. At the time of this writing in early 1997, the Convention had been ratified by 165 nations; the United States remained conspicuously absent from that long list.

During the CBD planning process[1], it was quickly recognized that an umbrella convention combining or consolidating existing conventions would be legally or technically impossible. Rather, the focus was placed on a framework treaty complementing existing treaties. Subsequently, it became apparent that many countries would not accept the narrowest concept of conservation to include only the preservation of wild resources. In response, the scope was expanded to include preservation as well as sustainable use, *ex situ* as well as *in situ* conservation, and access to commercial technologies as well as genetic resources. In short, the Convention grew to become a complete, if complex, framework treaty.

---

[1] For an overview of the planning process leading up to the adoption of the CBD, see Glowka *et al.* (1994), Introduction.

At the same time, it was evident that many countries, notably the G 77 nations (a non-aligned group of developing countries), resisted international mechanisms limiting the use of genetic resources under national jurisdiction. One concern was that the Convention framework would be used as a mechanism to compel developing countries to conserve, at national expense, biodiversity for international benefit. Consequently, much of the focus of the Convention is on national action, national plans, national sovereignty, and compatibility with national legislation. Placing so much of the operational aspects of the Convention in national hands of course risked that the harmonizing and gap-filling goals of the framework treaty would not be achieved, concerns which the hesitant outcomes of the first three Conferences of the Parties have, for some, reinforced. Here, the focus is on general economic incentives and approaches which, if adopted by multiple countries each acting in their own self interest, will contribute a degree of harmonization to the Convention process.

When the formal negotiating process began in February 1991, the Intergovernmental Negotiating Committee of the Convention on Biological Diversity was divided into two working groups. Working Group I dealt with general issues, including principles, obligations and conservation. Working Group II was given responsibility for access to genetic resources, technology access, technology transfer and the funding mechanism. This book maintains much of that division of topics by focusing on most of the Working Group II issues.

Finally, the emphasis on national actions necessitated the inclusion of a number of 'qualifiers' to individual country commitments under the Convention. Such conditional clauses as:

'develop, where necessary' (Article 8)
'as far as possible and appropriate' (Article 9)
'integrate ... into national decision making' (Article 10)
'endeavour to create conditions to facilitate' (Article 15)
'do not cause significant damage to the environment' (Article 16)

can be interpreted as a need to avoid mandating actions which exceed the means of many countries to carry out. Yet it is just those qualifications which in part make the interpretation of the Convention so complex, especially as the requirements are stated within national contexts where they may be viewed quite differently internally and externally.

The planning process led to three distinct if interrelated Convention objectives of conservation, use and benefit sharing (equity) (Article 1; emphasis added):

The objectives of this Convention ... are the *conservation of biological diversity*, the *sustainable use of its components* and the *fair and equitable sharing of the benefits* arising out of the utilization of genetic resources, including the *appropriate access* to genetic resources and by *appropriate transfer of relevant technologies* ...

In this volume, emphasis is on access to and use of genetic resources with benefit sharing, subjects which have a significant economic/incentives/ property rights component. That is, this work uses economic theory and practice as a pragmatic perspective from which to identify possible mechanisms at the national and local levels for achieving the Convention's objectives in the identified areas of access, use and benefit sharing. Conservation matters are not addressed directly, beyond the general expectation that what is perceived as having economic value tends to be conserved. Owing to the focus on access and use, attention here is directed more to the constitutional parts of biological diversity, genetic resources, than to ecosystems and habitats.

As definitions of genetic resources, genetic materials and biological diversity, those appearing in the Convention (Article 2) are utilized:

> *Genetic resources*: means genetic material of actual or potential value.
> *Genetic material*: means any material of plant, animal, microbial, or other origin containing functional units of heredity.
> *Biological diversity*: means the variability among living organisms from all sources including, *inter alia*, terrestrial, marine, and other aquatic ecosystems and the ecological complexes of which they are part; this includes diversity within species, between species, and of ecosystems.
> *Sustainable use*: means the use of components of biological diversity in a way and at a rate that does not lead to the long-run decline of biological diversity, thereby maintaining its potential to meet the needs and aspirations of present and future generations.

The treatment of genetic resources as the subset of genetic materials with actual or potential market value is particularly relevant here as emphasis is on the use of biodiversity, and, where needed, the creation of markets, to achieve relevant Convention objectives. This does not imply that the end goal of the Convention objectives is monetary – nor that funds can achieve all its objectives. Instead, it says that some objectives can be enhanced through market activities, including payment, and those are of particular focus here.

Note should be taken that nowhere in the Convention is 'conservation' defined. Conservation can be construed narrowly to mean protection or broadly to incorporate sustainable use. While the Convention text distinguishes between the two terms, in practice, conservation is sometimes used in its narrow and sometimes in its broad sense. For this volume, sustainable use is emphasized.

Principal Use and Sharing Articles of the Convention on Biological Diversity

As the Convention demarcates a complex agenda, it is of no real surprise that little specific has been accomplished through the initial three annual Conferences of the Parties (COP1, 2 and 3). The full depth of the agenda's complexity is evident from the text of the relevant use and sharing articles:

Article 8: In situ *conservation*

Each Contracting Party shall, as far as possible and as appropriate:

(a) Establish a system of protected areas or areas where special measures need to be taken to conserve biological diversity.
(b) Develop, where necessary, guidelines for the selection, establishment, and management of protected areas or areas where special measures need to be taken to conserve biological diversity.
(c) Regulate or manage biological resources important for the conservation of biological diversity whether within or outside protected areas, with a view to ensuring their conservation and sustainable use.
(i) Endeavour to provide the conditions needed for compatibility between present uses and the conservation of biological diversity and the sustainable use of its components.
(j) Subject to its national legislation, respect, preserve, and maintain knowledge, innovations, and practices of indigenous and local communities embodying traditional lifestyles relevant for the conservation and sustainable use of biological diversity and promote their wider application with the approval and involvement of the holders of such knowledge, innovations, and practices, and encourage the equitable sharing of the benefits arising from the utilization of such knowledge, innovations, and practices.

Article 9: Ex situ *conservation*

Each Contracting Party shall, as far as possible and as appropriate, and predominantly for the purpose of complementing *in situ* measures:

(a) Adopt measures for the *ex situ* conservation of components of biological diversity, preferably in the country of origin of such components.
(b) Establish and maintain facilities for *ex situ* conservation of and research on plants, animals, and microorganisms, preferably in the country of origin of genetic resources.

(c) Adopt measures for the recovery and rehabilitation of threatened species and for their reintroduction into their natural habitats under appropriate conditions.

(d) Regulate and manage collection of biological resources from natural habitats for *ex situ* conservation purposes so as not to threaten ecosystems and *in situ* populations of species, except where special temporary *ex situ* measures are required under subparagraph (c) above; and

(e) cooperate in providing financial and other support for *ex situ* conservation outlined in subparagraphs (a) to (d) above and in the establishment and maintenance of *ex situ* conservation facilities in developing countries.

Article 15: Access to genetic resources

1. Recognizing the sovereign rights of States over their natural resources, the authority to determine access to genetic resources rests with the national governments and is subject to national legislation.

2. Each Contracting Party shall endeavour to create conditions to facilitate access to genetic resources for environmentally sound uses by other Contracting Parties and not to impose restrictions that run counter to the objectives of this Convention.

3. For the purpose of this Convention, the genetic resources being provided by a Contracting Party, as referred to in this Article and Articles 16 and 19, are only those that are provided by Contracting Parties that are countries of origin of such resources or by the Parties that have acquired the genetic resources in accordance with this Convention.

4. Access, where granted, shall be on mutually agreed terms and subject to the provisions of this Article.

5. Access to genetic resources shall be subject to prior informed consent of the Contracting Party providing such resources, unless otherwise determined by that Party.

6. Each Contracting Party shall endeavour to develop and carry out scientific research based on genetic resources provided by other Contracting Parties with the full participation of, and where possible in, such Contracting Parties.

7. Each Contracting Party shall take legislative, administrative, or policy measures, as appropriate, and in accordance with Articles 16 and 19 and, where necessary, through the financial mechanism established by Articles 20 and 21 with the aim of sharing in a fair and equitable way the results of research and development and the benefits arising from the commercial and other utilization of genetic resources with the Contracting Party providing such resources. Such sharing shall be upon mutually agreed terms.

Article 16: Access to and transfer of technology

1. Each Contracting Party, recognizing that technology includes biotechnology, and that both access to and transfer of technology among Contracting Parties are essential elements for the attainment of the objectives of this Convention, undertakes subject to the provisions of this Article to provide and/or facilitate access for and transfer to other Contracting Parties of technologies that are relevant to the conservation and sustainable use of biological diversity or make use of genetic resources and do not cause significant damage to the environment.

2. Access to and transfer of technology referred to in paragraph 1 above to developing countries shall be provided and/or facilitated under fair and most favourable terms, including on concessional and preferential terms where mutually agreed, and, where necessary, in accordance with the financial mechanism established by Articles 20 and 21. In the case of technology subject to patents and other intellectual property rights, such access and transfer shall be provided on terms which recognize and are consistent with the adequate and effective protection of intellectual property rights. The application of this paragraph shall be consistent with paragraphs 3, 4 and 5 below.

3. Each Contracting Party shall take legislative, administrative, or policy measures, as appropriate, with the aim that Contracting Parties, in particular those that are developing countries, which provide genetic resources, are provided access to and transfer of technology which makes use of those resources, on mutually agreed terms, including technology protected by patents and other intellectual property rights, where necessary, through the provisions of Articles 20 and 21 and in accordance with international law and consistent with paragraphs 4 and 5 below.

4. Each Contracting Party shall take legislative, administrative, or policy measures, as appropriate, with the aim that the private sector facilitates access to joint development and transfer of technology referred to in paragraph 1 above for the benefit of both governmental institutions and the private sector of developing countries, and in this regard shall abide by the obligations included in paragraphs 1, 2 and 3 above.

5. The Contracting Parties, recognizing that patents and other intellectual property rights may have an influence on the implementation of this Convention, shall cooperate in this regard subject to national legislation and international law in order to ensure that such rights are supportive of and do not run counter to its objectives.

Article 18: Technical and scientific cooperation

4. The Contracting Parties shall, in accordance with national legislation and policies, encourage and develop methods of cooperation for the development and use of technologies, including indigenous and traditional technologies, in pursuance of the objectives of this Convention. [...]
5. The Contracting Parties shall, subject to mutual agreement, promote the establishment of joint research programmes and joint ventures for the development of technologies relevant to the objectives of this Convention.

Article 19: Handling of biotechnology and distribution of its benefits

1. Each Contracting Party shall take legislative, administrative, or policy measures, as appropriate, to provide for the effective participation in biotechnological research activities by those Contracting Parties, especially developing countries, which provide the genetic resources for such research, and where feasible in such Contracting Parties.
2. Each Contracting Party shall take all practicable measures to promote and advance priority access on a fair and equitable basis by Contracting Parties, especially developing countries, to the results and benefits arising from biotechnologies based upon genetic resources provided by those Contracting Parties. Such access shall be on mutually agreed terms.
3. The Parties shall consider the need for and modalities of a protocol setting out appropriate procedures, including, in particular, advance informed agreement, in the field of the safe transfer, handling, and use of any living modified organism resulting from biotechnology that may have adverse effect on the conservation and sustainable use of biological diversity.
4. Each Contracting Party shall, directly or by requiring any natural or legal person under its jurisdiction providing the organisms referred to in paragraph 3 above, provide any available information about the use and safety regulations required by that Contracting Party in handling such organisms, as well as any available information on the potential adverse impact of the specific organisms concerned to the Contracting Party into which those organisms are to be introduced.

Questions of Interpretation

The cited Convention articles contain difficult interpretative phrases related to access, use and sharing, including:

> 'equitable sharing of the benefits arising from the utilization of such knowledge, innovations, and practices' (8j)
> 'sovereign rights of States over their natural resources' (15)
> 'create conditions to facilitate access to genetic resources ... and not to impose restrictions that run counter to the objectives' (15)
> 'access ... subject to prior informed consent' (15)
> 'sharing in a fair and equitable way the ... benefits arising from the commercial and other utilization of genetic resources' (15)
> 'mutually agreed terms' (15 and 16)
> 'provide and/or facilitate access' (16)
> 'under fair and most favourable terms' (16)
> 'indigenous and traditional technologies' (18)
> 'priority access on a fair and equitable basis' (19)

These points emphasize that use need not only be efficacious and sustainable in achieving the goals of the Convention; it must also satisfy broad equity and ethical goals. Beyond that obvious interpretation, a clear understanding of the intent of the Convention framers has been elusive. Indeed, the IUCN Environmental Law Centre (Glowka *et al.*, 1994) prepared an entire document on interpreting the Convention language.

The text of Agenda 21, a 500-page blueprint detailing the 'new global partnership for sustainable development' adopted at the 1992 Earth Summit, provides additional insights into several relevant respects of the Convention (see Hunter *et al.*, 1994). Part 15.4(g), Conservation of Biological Diversity, calls for the recognition and fostering of traditional methods and knowledge, a point reiterated in 15.5(e). Part 34.3 expands the concept of relevant technologies to include 'not just individual technologies, but total systems which include know-how, procedures, goods and services, and equipment as well as organizational and managerial procedures'. Thus, human resource development and capacity building are an integral part of technology transfer.

More particularly, new and efficient technologies are described therein as 'essential to ... achieve sustainable development ...' (34.5). International business is specifically identified as 'an important vehicle for technology transfer' while emphasizing the need to incorporate business into the 'local or national culture' (34.11, 34.13). Finally, sustainable and efficient use of new technologies requires countries to 'build up technology assessment capacity' to include 'environmental impact and risk assessment' (34.26). These segments underscore the relevance the Convention drafters placed on matters of use, access, and equity.

An additional source of interpretive insights is gained through reviewing similar language used in conventions pre-dating the CBD. The Basel Convention (Transboundary Movements of Hazardous Wastes and Their Disposal) uses the term 'Prior Informed Consent' (PIC) (see Krummer, 1994). The requirements, however, are quite explicit; each signatory must identify a national authority for administering the procedures (Articles 6 and 7). Additionally, the information to be provided is specified in the Convention (Annex V A). Perhaps most significantly, the definitions of what is considered hazardous are quite explicit; under the CBD, the forms, the responsible parties, and the conditions triggering PIC are not always clear. Hence, the Basel Convention is far more explicit on the matter of PIC than is the CBD, meaning there is limited guidance to be found there.

The first specific international declaration of sovereign rights over natural resources was made in a UN General Assembly declaration of 1962 and subsequently reaffirmed for cultural property (G.A. Resolution 1803 (XVII), 14 December 1962; see Hunter *et al.*, 1994). The right, however, is not absolute; in the case of the CBD it is conditioned on efforts not to 'cause damage to the environment of other States'. Hence the CBD, while following general practices, does include unique aspects which can change the interpretation.

Objectives

The *interpretation* of the Convention on Biological Diversity remains, as noted, highly uncertain and open to multiple opinions. This is a significant issue in its own right, but it also has a direct effect on the Convention's *implementation*. In practice, interpretation and implementation will likely co-evolve, mimicking the species the Convention is intended to protect. There are several reasons to anticipate such a course of events. Many countries and local groups will continue to operate outside as well as within the Convention, meaning the Parties will have only partial control over the range of issues within the Convention's scope. In addition, many issues will not be further clarified until approaches are tested and the consequences observed. For example, the conservation value of sharing benefits with local communities was recognized in part through an assessment of successful projects: experience directed policy (see McNeely, 1988).

In this volume, emphasis is on means of implementation of the Convention on Biological Diversity. This is done with the full recognition that biodiversity and the human systems which use and protect it are changing rapidly; we no longer have the luxury of time for overall planning prior to action. Moreover, in many cases it is not clear what approaches will and will not be effective; that must be identified in multiple small-scale trials. All that can reasonably be expected at this juncture is the identification of the most promising directions for the beginnings. It is to this purpose that this book is directed.

Such a goal still leaves too broad a scope for making a focused contribution. My effort is delimited by placing emphasis on economic incentives for sustainable use of genetic resources. The focus on incentives is adopted in the spirit of the recognition that users should pay the full value for the resources they use (or misuse). Payment, however, implies in most cases that a market mechanism exists for the valuation and transfer of use rights. The absence of such mechanisms is one of the limiting factors in the appropriate use of genetic resources. Until recently, genetic resources had no major commercial value so there was no need for a market for their valuation and transfer. With the advent of biotechnology (and with it a mechanism for screening and transferring genetic materials), intellectual property rights (which in some cases allow for the claiming of ownership of applied knowledge) and the CBD (which enhanced ownership rights to the materials themselves), the conditions are in place for the establishment of efficient and equitable market mechanisms.

The creation of operational exchange mechanisms involves shared understandings of value, legal structures defining the extent of ownership and use rights, as well as the institutional market systems in themselves. Exploring each of these aspects is the objective of this book. This purpose will be served if some progress is made on a broader understanding of what is involved in the sustainable use of genetic resources. Such use can only be voluntary, by mutual consent, an understanding which cannot occur until the parties in transactions are informed of options and likely consequences. Setting out the options and consequences in as clear, complete and logical a manner as I can manage is the task before me now.

Chapter 2

Access Issues Under Article 15

Article 15(1) recognizes the 'sovereign rights of States over their natural resources' such that 'the authority to determine access to genetic resources rests with the national governments and is subject to national legislation'. Access issues have been on the agenda of all the COPs, and will continue at least to COP4. In this chapter, two general mechanisms for asserting sovereign rights over genetic resources are assessed – intellectual property rights (IPR) and contracts, the latter referred to here as Material Transfer Agreements (MTA). IPR is the subject of Chapters 7 and 9 as well where, respectively, their relationships with traditional knowledge and technology access are evaluated.

Prior to proceeding, it is important to note that sovereign rights over genetic resources are not a novel concept, for international practice has long recognized sovereign control over exhaustible resources like minerals and oil found within national borders (UN, 1974). It is also noteworthy that, while Article 15 does declare genetic resources to be governmental property, the delineation of who owns and controls genetic resources is a national legislative issue, just as with exhaustible resources. Governments must act formally to assert sovereign rights through national access legislation. Some have argued that national legislation is not required, that the reference to sovereign rights in the CBD (UNEP/CBD/COP/2/17), once ratified nationally, creates a legal requirement for benefit sharing, etc. That interpretation, however, is incorrect. The CBD establishes only national authority; domestic legislation is required to implement it (FAO, 1995; App. 3). Current access laws are described and examined in Chapter 3.

National sovereignty is delimited in several key regards. First, the Convention in this and other aspects applies only to activities, in this instance access, occurring after the CBD entered into force in December 1993

$(15(3))^1$. Typically, and understandably, international conventions are not applied retroactively (Glowka *et al.*, 1994), although this, in particular, leaves a significant quantity of agricultural genetic resources held in genebanks outside the scope of the Convention (see Chapter 6). Second, access shall be on 'mutually agreed terms' and 'subject to prior informed consent' with the 'sharing in a fair and equitable way ... the benefits arising from the commercial and other utilization of genetic resources' with the source country (15(4), (5) and (7)). Finally, access must be facilitated so as not to 'impose restrictions that run counter to the objectives of this Convention'. Issues of prior informed consent are treated in Chapter 5, and equity in Chapter 4. As background, we begin with a brief overview of *de facto* access systems up to the CBD.

Genetic Resources as a Common Heritage

There is no question that genetic resources have been widely shared. In a very basic sense, the origins of many common food crops are traceable to the 12 Vavilov centres of origin and over time have been dispersed worldwide. Presently, no region is completely self-sufficient in terms of origin of staple foods. The greatest self-sufficiency is two-thirds in the Indo-Chinese region (an origin for rice), and it is down to zero for North America and Australia (Table 2.1). This means the United States, a major food crop producer, is completely dependent on foreign germplasm, including potatoes from Latin America, corn from Central America, soybean and rice from China, and wheat from Syria and environs.

For industrial crops, the situation is even more pronounced. No region is even 50% self-sufficient, and most are 75% dependent, or more (Table 2.2). North America posts an 85% dependency, lessened only by the existence of sunflower. Livestock origins are not as commonly documented, but for one fauna – the nearly ubiquitous chicken is of Asian origin. Outside agriculture, an estimated one quarter of processed medicines originate from plants (Principe, 1988). Here, as well, the sources are quite dispersed, including digitalis/foxglove/Europe, penicillin/moulds/North America, and cancer/rosy periwinkle/Madagascar.

Much of this transfer pre-dates recorded history, and could date back well into the 10,000–12,000 years of species domestication and food

---

[1] The exact wording is 'acquired ... in accordance with this Convention'. A second stipulation for Convention authority, connected by an 'or' suggesting equal standing, is 'provided by Contracting Parties that are countries of origin of such resources'. One interpretation of this clause is the extension of the Convention requirements to *all* genetic resources from centres of origin of those materials regardless of the date of transfer. That possibility has not been explored in the literature and does not appear to be an active issue at this time.

Table 2.1. Regional sources and utilization of major food crops*.

Regions of production	Regions of diversity											Total dependence
	Chino-Japanese	Indo-Chinese	Australian	Hindustanean	West Central Asiatic	Mediterranean	African	Euro-Siberian	Latin American	North American		
Chino-Japanese	37.2	0	0	0	16.4	2.3	3.1	0.3	40.7	0	=100%[†]	62.8
Indo-Chinese	0.9	66.8	0	0	0	0	0.2	0	31.9	0	=100%	33.2
Australian	1.7	0.9	0	0.5	82.1	0.3	2.9	7	4.6	0	=100%	100
Hindustanean	0.8	4.5	0	51.4	18.8	0.2	12.8	0	11.5	0	=100%	48.6
West Central Asiatic	4.9	3.2	0	3	69.2	0.7	1.2	0.8	17	0	=100%	30.8
Mediterranean	8.5	1.4	0	0.9	46.4	1.8	0.7	1.2	39	0	=100%	98.2
African	2.4	22.3	0	1.5	4.9	0.3	12.3	0.1	56.3	0	=100%	87.7
Euro-Siberian	0.4	0.1	0	0.1	51.7	2.6	0.4	9.2	35.5	0	=100%	90.8
Latin American	18.7	12.5	0	2.3	13.3	0.4	7.8	0.5	44.4	0	=100%	55.6
North American	15.8	0.4	0	0.4	36.1	0.5	3.6	2.8	40.3	0	=100%	100
World	12.9	7.5	0	5.7	30	1.4	4	2.9	35.6	0	=100%	

* Reading the table horizontally along rows, the figures can be interpreted as measures of the extent to which a given region of production depends upon each of the regions of diversity. The column labelled 'Total dependence' shows the percentage of production for a given region that is accounted for by crops associated with non-indigenous regions of diversity.

[†] Because of rounding error, the sum of the figures in each row does not always equal 100.

Source: Kloppenburg and Kleinman (1987).

Table 2.2. Regional sources and utilization of major industrial crops*.

Regions of production	Regions of diversity											Total dependence
	Chino-Japanese	Indo-Chinese	Australian	Hindustanean	West Central Asiatic	Mediterranean	African	Euro-Siberian	Latin American	North American		
Chino-Japanese	8.3	4.7	0	1.4	7.4	27.5	0.1	0	45.4	5.1	=100%[†]	91.6
Indo-Chinese	5	43.5	0	7.1	2.9	0	22.6	0	18.8	0	=100%	56.4
Australian	0	51.2	0	0	1.8	3.3	0	0	15.4	28.3	=100%	100
Hindustanean	2.6	14.2	0	7.2	20.5	17.2	0.9	0	35.2	2.1	=100%	92.7
West Central Asiatic	1.5	14.7	0	0	4.5	14.2	0.1	0	56.6	8.4	=100%	95.5
Mediterranean	0	3.9	0	0.2	2.4	25.3	0	0	31.8	36.5	=100%	74.9
African	1.3	16.3	0	0.1	10.6	0.4	22.4	0	46	3	=100%	77.7
Euro-Siberian	0.4	0	0	0.1	12.8	41.3	0	0	17.5	27.9	=100%	100
Latin American	0.2	30.4	0	0.4	5.9	0.4	25.7	0	28	9.1	=100%	72.1
North American	0	3.7	0	0	8.3	33.1	0	0	39.6	15.3	=100%	84.7
World	2.1	13.7	0	2	10.8	18.2	8.3	0	34.4	10.5	=100%	

* Reading the table horizontally along rows, the figures can be interpreted as measures of the extent to which a given region of production depends upon each of the regions of diversity. The column labelled 'Total dependence' shows the percentage of production for a given region that is accounted for by crops associated with non-indigenous regions of diversity.

[†] Because of rounding error, the sum of the figures in each row does not always equal 100.

Source: Kloppenburg and Kleinman (1987).

production. The treatment of ownership during that period is similarly unknown. Systematic, documented exchange coincides with the period of European exploration and establishment of the plantation system. This period, called the 'Colombian exchange', began in 1494 with the transportation of sugarcane to Hispaniola on Columbus' second voyage. Sugarcane, with origins in Southeast Asia, travelled through Syria and Egypt to the Mediterranean *en route* to the New World. The plantation system, in turn, engendered the slave trade and the need to feed cheaply large numbers of non-native peoples. Captain Bligh of the *Bounty* was returning from Tahiti in 1789 with a load of breadfruit plants when the ill-fated mutiny took place. Less well known is the successful return trip which led to establishing breadfruit in the West Indies in 1793.

The most systematic efforts of exchange involved botanic gardens in national attempts to monopolize commodities or, alternatively, to break another government's monopoly (Brockway, 1988). A single coffee tree reaching the Amsterdam botanic gardens in 1706 from Ethiopia via Ceylon and Java became the basis for the New World coffee industry. Today, botanical gardens are particularly associated with tree species, which were initially collected in part as wood sources for shipbuilding following the near denuding of European forests by the sixteenth century. One basis of England's interest in colonizing India was the extensive teak forests there (McNeely *et al.*, 1995).

During the colonization period, plant genetic resources were treated as one of the spoils of occupation, protected by military strength. When access could not be gained through superior strength, stealth was employed. Both were involved in the spice trade, long dominated by the Dutch from Java. In 1755 a young Frenchman, Pierre Poivre, smuggled pepper and cinnamon to Mauritius (then the Ile de France) where it fell into British hands when the islands were captured during the Napoleonic Wars. This was a sovereign rights system of sorts, but one buttressed by military power as opposed to legal processes.

With the decline of the empire system into the eighteenth century, governments lacked the military presence and legal authority to compel sovereign nations to yield valuable germplasm. Outright theft was sometimes substituted. In a celebrated case, Henry Wickham is credited with making off with 70,000 rubber plant seeds from Brazil in 1876. Dramatic accounts of that endeavour appear to stem from Wickham's self-engrandisement. In reality, there were at the time no prohibitions on the export of rubber seeds and the seeds were likely collected 'almost certainly with the knowledge of authorities' (Hecht and Cockburn, 1989, pp. 90–92). Travelling through Kew Gardens and Singapore, these seeds provided the basis for the Malaysian plantation rubber industry, now the leading source of natural rubber. Horticulture was not exempt from these commercial interests, as exemplified by the Dutch 'tulipmania' of 1634–37 when some breeder bulbs reached prices in the $1000s, high now and a fortune at the time, before collapsing (Gaber, 1989).

Note should be taken that these efforts were focused on commercial crops, not staple foods. Germplasm exchange of food crops is less well documented in part as these resources seemed to be treated as community property open to all, or alternatively, as personal property to be dispersed worldwide at will. Clearly, much of the seed (and livestock breeds) used in the Americas and Oceania arrived with European settlers. Hard red winter wheat, for example, has been traced to immigrant Mennonite farmers from the Ukraine although the key dwarfing gene originated in Japan (Brush, 1996a, p. 144). Seed appears to have been saved on individual farms or traded within communities, later in conjunction with cooperatives. Sold seed was handled by merchants in a completely unregulated manner or by aristocratic landowners who became involved in seed improvement as part of a push to scientific farming in the late eighteenth century. By the mid-1800s, selection, cleaning of foreign matter, and pure lines were well-established concepts. Recent studies of on-farm selection among traditional farmers indicates the concepts are indeed well understood (see, for example, Eyzaguirre and Iwanaga, 1996). In truth, prior to the advent of systematic plant breeding, seed sales were already established, but predominately on a local and small scale.

The basis for systematic plant breeding was developed by Mendel in the 1860s but was not recognized until the laws' rediscovery about 1900 (brief history in Busch *et al.*, 1991, Chap. 3). This had an immense impact on the practice of breeding, particularly after 1920, when trained breeders utilizing inbreeding and crossing thus had an advantage over farmers undertaking selection and varietal crossing (Fitzgerald, 1990, Chapter 1). Such a systematic approach, along with the Mendelian discovery of genes which were identified with useful traits, created a use for genetic materials not previously incorporated into varieties in widespread use.

This division of labour between the public (breeding) and private (propagation) sectors in developed countries changed first for hybrids, particularly corn, with early true hybrids appearing about 1920 (Fitzgerald, 1990). By the 1950s the major corn growing areas in the USA had converted 95% of their acreage to hybrids. Some prime areas were converted even earlier. Hybrids, unlike open pollinated crops, do not reproduce true to form and must be purchased for each crop. This provides a form of biological protection for breeders such that an investment in breeding could be justified by private firms. One of the early entrants, Pioneer Hi-Bred (established in 1926), is today the world's largest seed company in terms of sales. Today, public breeding of corn in the US is limited to some pure lines and commercial varieties in fringe producing areas.

In contrast to breeders of hybrid lines, private breeders of open pollinated seed had to wait until the development of intellectual property rights in the form of Plant Breeders' Rights (PBR) in 1960 to achieve legal restrictions on copying (see Chapter 9 for a discussion of PBR and other forms of IPR). For developing countries, due to a mix of government policy and market considerations, privatization in the seed sector has been far more recent (Cromwell *et al.*, 1992).

An early *ex situ* systematic collection of genetic materials for wheat breeding purposes was assembled by the US Department of Agriculture in the 1920s. It was, however, only in association with the establishment of the International Agricultural Research Centers (IARC), beginning with the International Rice Research Center in 1960, and the aftermath of the Green Revolution, that systematic collection and *ex situ* storage began on a worldwide basis (Wilkes, 1988) (see Chapter 6). Our concerns here are the agreements under which the materials had been collected.

In most instances, until the very recent past, no detailed agreements accompanied the germplasm collections. Rather, the process of free exchange was adhered to, particularly by the IARC, and typically took place directly between scientists with little government involvement. This open system facilitated exchange and is attributed with making a major contribution to the advances in food crop productivity under the Green Revolution (Barton and Siebeck, 1994). Informally, the materials were treated as the 'common heritage of mankind', but formally it was in 1983 with the Food and Agricultural Organization's (FAO) International Undertaking on Plant Genetic Resources when the label of 'a common heritage of mankind' was applied (Resolution 8/83 of the 22nd Session of the FAO Conference). While the impetus was *ex situ* collections, the implication of the resolution extended to all food crops. Other aspects of the Undertaking address conservation and preservation (see Chapter 6).

More recently still, in 1988, the conditions under which the genetic materials were transferred were clarified under the concept of 'in trust'. 'Trusteeship' is a form of holding material for the benefit of a third party, a beneficiary. Trustees can do almost anything with the subject matter entrusted to them so long as they do so in accordance with the understanding of those who entrusted the material to them and act in pursuance of the interest and to the benefit of the beneficiary' (Barton and Siebeck, 1994).

In economic terms, property rights are typically created in response to rises in value. According to Barzel (1989, p. 65), '... rights in the sense of the ability to gain from property are largely a matter of economic value rather than of legal definition'. This factor can be seen in the extension to the continental shelf jurisdiction from 12 to 200 miles at a time when deep oil drilling was perfected and near-shore fish stocks began to decline. Similarly, the Law of the Sea was negotiated at a time when interest arose in mining mineral nodules from the ocean floor. Genetic resources are following a similar pattern of interest in explicit property rights at a time when biotechnology increases the value of those resources. Property rights, it can be demonstrated theoretically, are not required when knowledge is perfect and transactions are costless. Property rights, however, serve a clear economic role in the less perfect real world with its relatively high transaction costs.

But by the early 1980s broad allegiance to the common heritage concept had begun to unravel, in part due to PBR. Some developing country advocates complained they were donating their materials only to buy them back from wealthy multinational breeding companies (see, for example, Fowler, 1995). Breeders, for their part, argued that they were simply being compensated for their value added, noting further a general willingness to pay for input materials if a system existed. Truth applied on both sides, but clearly there was an imbalance in institutional arrangements. Subsequent modifications to the Undertaking can be seen in part as attempts to rebalance that situation. Principal is the concept of Farmers' Rights developed by the FAO under the so-called Revised Undertaking for Plant Genetic Resources (Resolution 5/89). While the Undertaking is not necessarily restricted to plants with agricultural applications, it is quite evident that is its intended focus. Farmers' Rights is not referenced directly in the CBD, but is in the Nairobi Final Act, a kind of preface to the Convention. Resolution 3(4) recognizes:

> the need to seek solutions to outstanding matters concerning plant genetic resources ..., in particular:
> (a) Access to *ex situ* collections not acquired in accordance with this Convention.
> (b) The question of Farmers' Rights.

In Resolution 5/89 Farmers' Rights are defined as 'rights arising from the past, present and future contributions of farmers in conserving, improving, and making available plant genetic resources ...'. Farmers' Rights are to be 'implemented through an international fund on plant genetic resources which will support plant genetic conservation and utilization programmes, particularly, but not exclusively, in the developing countries' (FAO Resolution 3/91, Annex 3 to the International Undertaking). The fund itself was initially agreed upon in 1987 at an estimated cost of $600 million (quoted in UNEP/CBD/IC/2/13, 1994, Par. 38). No further details on the implementation and operation of this fund were included, and to date it has not been implemented on an international basis. India, however, has planned for a national fund based on a levy on domestic seed sales (Damania, 1996).

At the same FAO session in 1989, a second resolution (4/89) was adopted which provided for an 'agreed interpretation' that PBR were not incompatible with the Undertaking as traditional farmers have no 'rights' to protected materials. Practically speaking, this means protected materials lie outside the Undertaking. Barton and Siebeck (1994) interpret this recognition of PBR as the quid pro quo for the endorsement of Farmers' Rights by developed countries, the first such rationale in the access debate. A final restriction on free exchange was adopted in 1991 (Resolution 3/91) when the common heritage principle was subordinated to 'the sovereignty of the states over their plant genetic resources' in accordance with the CBD then being drafted. The CBD itself can be seen in part as a wide endorsement of the cessation of open access in exchange for a system providing some remuneration. But what system or systems?

Approaches can conveniently be classified into three categories, as follows:

1. Unilateral – patents and other traditional and non-traditional forms of IPR.
2. Bilateral – contracts.
3. Multilateral – predominately open exchange systems, including the common heritage approach.

In the remainder of this chapter, the conceptual issues and general implications of the unilateral and bilateral approaches are explored. Proposed multilateral systems are the subject of Chapter 6.

Unilateral Systems/Intellectual Property Rights (IPR)

IPR is referred to as a unilateral system because it operates as national law; certain rights are granted to limit the options of non-IPR holders within a country. Patent holders, for example, can (depending on the particulars of national law) prohibit the use, production, marketing or importation of a product into countries where valid patents are held. Similar restrictions apply for varieties protected under PBR, which pertain to plants for agriculture and horticulture. Thus the systems are unilateral – national prohibitions which operate without the consent of other parties. By contrast, contracts are binding only on those who enter them. There are international IPR agreements and conventions, but these harmonize protection across national boundaries, they do not authorize it. The role and functions of traditional IPR, consisting, in the case of genetic resources, of patents, PBR, and trade secrets, are described in Chapter 9. Here their role in controlling access to genetic resources (and hence strengthening opportunities for benefits sharing) is evaluated. We begin first with the conventional forms, progressing to other less conventional systems.

The benefits of IPR to owners are immediately apparent. The holder of a patent or certificate of PBR can prohibit the importation, manufacture, or sale of products using the protected materials. Usefulness then hinges on: (i) whether IPR will provide meaningful opportunities to exclude unauthorized use for genetic resources, and (ii) the practicality of use.

Traditional intellectual property rights

Applicability of PBR

PBR, it can be stated categorically, are not broadly useful for the claims of ownership of genetic resources. The UPOV acts adopted by most countries with these systems[2] are clearly directed to agriculture and horticulture applications. This is most evident from the 1961 Act where (Annex) the families and genera included are within what can be characterized as agriculture and horticulture. The 1978 and 1991 Acts are not as specific, but an application for materials from the wild with possible medicinal uses would be treated very sceptically. Certainly the national PBR programmes are not equipped to evaluate the acts (for a description of the operation of one such programme see Lesser, 1987a). Hence PBR should be considered within the agricultural realm only.

Even with agricultural applications, however, the applicability of PBR is limited. PBR protects the entire plant, its propagating materials and, under the 1991 Act, marketed parts, such as flower blooms. What it does not protect is the distinguishing characteristic *per se*, say disease resistance. If that characteristic were conferred by a gene construct, there would be no control over the removal of that construct and insertion into an unrelated variety, or use in another form. A full justification for that conclusion under the 'essentially derived' option of the 1991 Act (Article 14(5)) is somewhat lengthy and detailed, so only a synopsis is provided here.

The 1991 Act in Article 14(5) allows for two categories of protection, for varieties which are 'essentially derived' and those which are 'initial' varieties. Essentially derived varieties are 'predominantly derived from the initial variety, or from a variety that is itself predominantly derived from the initial variety, while retaining the expression of the essential characteristics ...'. Permission is required from the holder of the initial variety (if itself protected by PBR) for the multiplication or sale of predominantly derived varieties (that is, commercial but not breeding use).

The complexity is in the working definition of 'essential derivation'. Actual implementation by a national PBR Committee has yet to occur so any definitions at this stage are preliminary. None the less, proposals have used terms like 'preponderance of genetic material' (see Hunter, 1992; Rasmussen, 1990). These suggest a numerical standard – the source of the

---

[2] UPOV is the French acronym for the International Union for the Protection of New Varieties of Plants, which supports the international harmonizing convention for PBR. In addition, a few countries like Kenya and Taiwan have national PBR systems, but the details are not readily available.

bulk of the genetic material – as opposed to the source of the agronomically important traits. Now since much of the genetic material in landraces, a possible application for PBR protection, is not suited to the requirements of commercial varieties and must be removed during the breeding process, a landrace is unlikely to become an initial variety under the UPOV 1991 Act. Moreover, there is to be but one initial variety for any essentially derived variety (UPOV, 1992). Taken together, the interpretations of essentially derived and single parent variety mean royalties will not be collectable from derivative varieties. That is in effect no protection for either the source landrace or the genetic coding for the useful trait.

Applicability of patents

Among traditional forms of IPR, genetic resources are most likely protectable by patents, if at all. There is, for example, no conceptual reason why genetic resources in the form of, say, plant materials (where permitted) identified from the wild cannot in concept be patented. Practice in the US requires there be human diligence (author's term) involved; the purification of a bacterial strain or the identification of a unique rose mutant are examples. Regarding the situation with genomic DNA sequences in Europe, Straus (1994, p. 20) writes, 'the prevailing view and the patent office practice is that human technical intervention is required to recognise them [DNA sequences] and to protect them in a reproducible manner. The subject matter of such application, therefore, is an invention and not a mere discovery'. The distinction between invention and discovery is key because discoveries are excluded from patent protection under the European Patent Convention (EPC Article 5(2(a))). What is inhibitory to the patentability of the bulk of materials identified from the wild is the 'utility' ('industrial application' in Europe) requirement for patenting (US Patent Act Section 101, EPC Article 52(1)). In simple terms, the utility requirement specifies a use of the invention be identified. The use need not be original (although the invention must be) nor practical in an efficiency sense, but there must be a specific use designation. The infamous patent applications under the Human Genome Project were withdrawn for failure to satisfy that requirement. The great bulk of materials found in the wild similarly lack a known use.

Even when that requirement can be met, the cost factor must be considered. US patent applications cost an estimated $20,000 (and about twice as much in Europe because of translation fees)[3] (Abbott, 1993). That is an implausible amount for protecting a large collection of materials the vast bulk of which will have no real market value.

---

[3] In 1996 the European Patent Office cut fees by 20% and delayed the payment of country designation fees by 18–24 months, but did not adjust translation requirements (Meller, 1997).

Applicability of trade secrets

Trade secrets require reasonable efforts be used to keep valuable information confidential. Examples of protectable information include the formula for Coca-Cola syrup, a customer list, or crossings used for a hybrid. Legislation typically allows for the recovery of financial losses if information is improperly acquired – such as if a former employee reveals corporate secrets to a competitor during a contractually specified period – but no recovery in the case of careless/accidental loss of control, or independent discovery. Open pollinated seeds, which by definition recreate themselves on planting, preclude the use of trade secrets, although trade secrets are applicable to hybrids provided the pure lines are carefully guarded. The refusal of local/indigenous communities to share their knowledge with outsiders is a form of trade secret protection. From a public perspective, trade secrets are not desirable as they limit the flow of information. Overall then, trade secrets have a limited applicability for protecting owners of genetic resources.

Non-traditional forms of IPR

If traditional IPR is largely inapplicable, what of non-traditional forms? This would include such systems as protection of folkloric expressions and appellations of origin, examined here. For more information see Lesser (1994a), Correa (1994), and Posey (1994).

Farmers' Rights

Farmers' Rights as noted above is the term developed by the FAO under the so called Revised Undertaking for Plant Genetic Resources. While not necessarily restricted to plants with agricultural applications, it is quite evident that is the intended focus of the Undertaking.

In concept, Farmers' Rights operate more as a moral obligation than an economic incentive. They are not connected with any specific future action but rather with a general conservation and equity objective. This is noted without prejudice but only to emphasize that the objectives, and hence the likely results, of the system are quite different from IPR.

Perhaps the major comment which can be made is the lack of contributions to the compensation fund since its proposal. The time span has been relatively short, but there have been few indications to date that such a fund will be constructed. The entire International Undertaking process received much negative attention in the developed countries early on due to the interpretation of 'plant genetic resources' to refer to both unimproved and improved genetic materials (Article 5) (see Grossman, 1988). Private firms have not made their products available without charge, and while it is a matter of interpretation if that was specifically required by the Undertaking, the notion that private firms could be required to give away their products poisoned the atmosphere and a fund never materialized. Subsequently, the proposed tax on seed sales was never supported.

It is, of course, possible that the concept of Farmers' Rights could be pursued more readily under a different name and institutional structure[4], possibly under the Biodiversity Convention. And possibly some of the promised financial resources under the Convention could be used for such a purpose. But there is little evidence to date that the major financial donor countries would support such a farmers' fund, or to what purposes the monies would be put.

Folklore

Many of the issues associated with protecting genetic materials have parallels in protecting expressions of folklore. That is particularly true of landraces which, like folkloric expressions, are the result of long-term community contributions. And, again as with landraces, there is no system of compensating, or even acknowledging, those communities for their contributions. The applicable IPR systems, copyright and trademark, operate similarly to patents in requiring new and unique creations, which folklore is not. Perhaps then attempts to protect folklore will provide some insights for use with genetic materials.

Treatments of IPR for folklore culminated in the joint (but unimplemented) 1985 'Model Provisions for National Laws' by WIPO and UNESCO (WIPO, 1985). There, the expressions of folklore are defined as 'characteristic elements of the traditional artistic heritage developed and maintained by a community ... or by individuals reflecting the traditional artistic expectations of such a community'. These expressions may be verbal (folk tales), musical or action (dances) as well as tangible expressions such as art, musical instruments and architectural forms (Model Law, Section 2). When used 'with gainful intent outside their traditional or customary context' such expressions are 'subject to authorization' by the competent authority of the community (Section 3). The expressions may originate from the community or elsewhere, provided they were subsequently further developed, adopted, or maintained through generations (Par. 35).

As can be seen, the issues are indeed similar to those for selected genetic materials like landraces. However, no helpful detail is included on how to implement what can only be described as concepts. For example, in the frequent situation where neighbouring communities practice slight variants of the same tradition, whose permission would be required – any one of the communities, or some/all of them? How or who would determine when an expression is different enough to be a separate form of expression? What competent authorities would be identified to represent a community?

---

[4] The author has previously proposed 'Plant Genetic Resources Foundation' or similar neutral-sounding name.

And what constitutes an 'artistic heritage'? Would 'cowboy boots' be considered as an artistic heritage? They are preserved by the culture, are important for personal and regional identity, and may be elaborately decorated. If so, would royalties for copying cowboy boots have to be paid to Texas, the USA Southwest, all North America, including Mexico and Canada? And what characterizes a cowboy boot: a shape, a kind of decoration, or exactly what? While an extreme example, these illustrate the kinds of decisions that would have to be made.

All IPR systems involve similar definitions of scope and identification of ownership. A full system, however, includes a definition of who makes the decisions (the national patent office) on what basis (the application, especially the patent claims) and interpretation of unauthorized use (making, selling). The intent is to make the process open and systematic so that it is possible to know within reasonable bounds (questions will always remain in individual cases) what is protected and what would be infringing. It is that kind of specificity which is lacking from this model law and from a system for genetic materials. An evaluation of efforts to develop a system for folklore helps clarify the issues but contributes little to the development of a system for genetic materials.

Codes of conduct

Codes of conduct refer to standardized but voluntary agreements specifying obligations. They are similar to a one-sided contract voluntarily entered (compare with, for example, Downes *et al.*, 1993). The FAO has over several years prepared a 'Code of Conduct for Plant Germplasm Collecting and Transfer', still in draft form, which could serve as a model for claiming ownership over genetic materials (FAO, 1993).

The Code, which is directed primarily to governments, has the principal objectives of promoting respect for the environment, for local traditions, and for cultures, and establishing mechanisms for compensating local communities and farmers for their conservation and development activities (Article 1). The mechanism for achieving these goals is to require collection permits (Article 8) subjectable to certain conditions, including 'financial obligations', restrictions placed on the distribution or use of the germplasm or improved materials derived from it, the use of care in the collecting process, and provision, on the country's request, of duplicate sets of collected materials (Articles 8, 10 and 11).

Separate obligations apply to sponsors ('see to degree possible collectors abide by Code', Article 12), curators (provision of further samples, Article 13) and users ('consider providing some form of compensation', Article 14). This Code is seen as serving temporarily until national legislation is passed, or possibly a legally binding international agreement like a protocol under the CBD. A protocol is a separate agreement, binding on those countries adopting it.

The Code is in full accord with the Convention as it provides a means by which the national sovereign rights can be implemented (Convention Article 3). Indeed, little will change at the national level until national governments establish controls through such mechanisms as this. Thus the Code can be seen in part as a model law for national governments. In its present form as a voluntary guideline it depends on the material users/ collectors for application.

Other codes have been directed toward use by collectors and hence structured somewhat differently. The detailed draft of a code of conduct used by the Royal Botanic Gardens at Kew (UK) is quite specific in demanding of its collectors 'best efforts' in securing informed consent at the national and local levels. Best efforts also apply to collaborations, benefit sharing, and capacity building. Benefit sharing is to be accomplished through the use of contracts, discussed below.

Cultural property

'Cultural properties' in the tangible forms of ancestral remains and artefacts have been successfully reclaimed from museums and researchers worldwide by indigenous groups (references in Posey, 1994). In some countries, groups have been successful in regaining title, or at least access, to areas of spiritual importance. With this basis, some groups have claimed these rights to 'cultural properties' in other forms. In the terms of the Matautu Declaration from the First International Conference on the Cultural and Intellectual Property Rights of Indigenous peoples (UN 1993 Commission on Human Rights, E/CN.4/Sub.2/AC.4/1993/CRP.5.26 July 1993) (preamble, emphasis added):

> We declare that Indigenous peoples of the world have the right to self determination; and in exercising that right must be recognized as the *exclusive owners of their cultural and intellectual property.*

While 'cultural and intellectual property' are not defined, reference is made to ownership by 'those who created them' and 'peoples who have inherited such knowledge' (Articles 2.4 and 2.7). Like Farmers' Rights, this too is more of an appeal to moral responsibility than an operable ownership system. Lacking are vested rights and operational definitions. The significance of identifying owners of traditional knowledge becomes evident in the application of the Philippines access law described in Chapter 3.

From the preceding it is evident there indeed is an imbalance in the effects of IPR, both conventional and non-conventional, regarding 'western' and traditional products. The call has been for a new system, especially regarding the intellectual property of local and indigenous peoples (see, for example, Posey, 1994). This subject is addressed in Chapter 7. For the present it is evident that existing unilateral systems are inadequate for *directly* controlling access and use of genetic resources. Proposals have been made for indirect use of IPR as a form of enforcement mechanism to protect rights in genetic resources. These approaches are considered next.

Indirect uses of IPR

Patents and PBR are useful only as long as they remain in effect, making revocation a significant threat. This underlies some recent proposals for utilizing IPR to protect genetic resources rights, often for local and indigenous communities, but the approaches are general as to the owners (e.g. Yamin, 1995; literature review in UNEP/CBD/SBSTTA/2/7, 1996).

The general concept is as follows. Patent applications require a description of the invention (see Chapter 9), frequently containing an indication of sources for the materials used. PBR also require some description of the source lines. The proposal is to use these descriptions to enforce the rights of the owners of those source materials. The particulars of how that might be done vary. One proposal would require the patent, PBR, or other office to verify that identified materials were legally acquired (UNEP/CBD/SBSTTA/2/7, 1996, Par. 92; examples of current practice in UNEP/CBD/COP/3/23, 1996, Annex 1). Less formally, the office would contact the relevant owners who would be responsible to challenge the patent if access was unauthorized. Both approaches would require some change in patent legislation, or at minimum some change in administrative practices.

While relying on a patent (or Plant Breeders' Rights) office to enforce ownership rights may be effective in some specific instances, it is unlikely to be effective as a general approach. Limitations include the following (OECD, 1996):

1. Only a portion of commercialized genetic resources is ever patented. Anything not patented would be unaffected by the policy.
2. Also possibly unaffected are useful compounds suggested by the natural materials but not directly derived from them. A patent office does not have the stature to determine ownership of such derivatives.
3. Patent legislation is slow to amend, making any necessary changes a protracted process. Moreover, practitioners typically resist expanding the role of the patent office beyond the evaluation of applications, feeling that other issues are beyond their competence. This means objections to enhancing the roles of those offices can be anticipated.
4. Countries will still require access legislation, or similar accords, which establish the legal ownership basis on which the patent office is asked to operate. A more direct approach to controlling unauthorized use would be through access legislation instead of the present patent system.

Overall, these proposals appear to have limited relevance and practicality at this time. They may serve in some instances, but by all means should not be seen as a general resolution. Basically, under prevailing practices, it is the owner's responsibility to monitor use and misuse and bring suit against unauthorized use. Expecting an institution to act on an individual's (or community's) behalf is unrealistic.

Bilateral Systems/Material Transfer Agreements

Bilateral systems refer to direct, two party agreements between suppliers and users. They are best described as contracts or, applied to genetic resources, as Material Transfer Agreements (MTA). MTAs are being used when the owners of the materials are known and willing to provide use permission at least conditionally. The roles and functions of these agreements are described in this chapter. MTAs, however, are of limited use when the ownership of the material has not been clearly established, such as with much material found in the wild and even with, say, landraces, the seeds of which are sold in local markets as food. Furthermore, most countries presently do not prevent the removal/use of genetic resources, meaning that the negotiation of an MTA is quasi optional. Access legislation, the subject of Chapter 3, serves the function of establishing general terms/limitations of access, thereby establishing the requirement for negotiating an MTA.

Simple material transfer agreements

A simple MTA (see Box 2.1) would allow research use only and contain three principal aspects:

1. A description of the material.
2. A statement that only research use is permitted; commercialization would require an additional agreement.
3. An agreement prohibiting distribution to third parties.

These agreements are typically signed and returned prior to the distribution of the materials. A single agreement may incorporate multiple related materials. Finally, given the potential legal liability resulting from use of the materials, language will often be added to the effect that no warranties of usefulness or appropriateness are implied, nor are there warranties that commercialization of a resultant product would not infringe another patent. A biotech product, for example, may have had employed in its development a proprietary vector or technical procedure so that permission from the patent holder(s) will be required prior to commercialization. The license is typically made legally responsible for identifying legal claims, and, when needed, securing permission. In short, users must beware.

Such agreements have become quite routine, and, while a standard agreement has yet to emerge, the Association of University Technology

Box 2.1. Clauses of a simple material transfer agreement.

RE Biological Material identified as ....

1. *Recipient* may only use the Biological Material for research purposes and no derivatives of the Biological Material may be distributed to any third party.

2. *Recipient* may not distribute the Biological Material provided herein to any other party.

3. The Biological Material and any derivatives thereof may not be commercialized without a license from *Foundation.*

4. It is expressly understood that no right to a license is given or implied by this agreement.

Source: Cornell Research Foundation, Inc.

Managers (USA) has proposed a series of definitions which are receiving widespread use. Definitions include those of derivatives, divided between unmodified and modified. Unmodified derivatives include 'unmodified subunits' while modified derivatives must 'contain/incorporate' the provided material. Typically, the commercial benefit sharing agreement is not negotiated until and unless a use has been found. This places the user in some jeopardy as considerable resources may have been invested by that point, with no promise of an equitable agreement emerging. As a counterbalance, the owner of the material has an incentive to facilitate commercialization as a source of revenue.

Administratively, these agreements are straightforward. Scientists have the additional responsibility of maintaining records and identifying use restrictions prior to commercialization or further distribution. Noting use restrictions is especially an issue for partially developed materials requested by a colleague because they are often embodied in the making of a commercial product so it is easy to lose the identity of one constituent part. Research scientists often describe these administrative matters as a nuisance but not a serious impediment. The records are maintained anyway; the task is to use them in an as-yet uncommon way.

Knowledge of infringement offences is indirect, say through reports from competitors of commercial propagators, as opposed to a systematic effort to track materials through records. Genetic fingerprinting is technically feasible for well-described, stable materials (such as commercial varieties), but would cost upwards of $170–500 per variety. For wild plants and many landraces which are neither homogenous nor characterized, costs could be 10–20 times that due to the need for marking more base pairs in several replicates. Those costs quickly become untenable when the limited commercial value of most genetic material is recognized (Lesser, 1997; see also Chapter 6, Table 6.3).

Consequences of material transfer agreements

An additional matter of concern is the effects of MTAs on the exchange of materials and the general flow of information, particularly for the public sector. A general expectation is that MTAs will prohibit/delay exchange, or at a minimum will formalize the arrangements among researchers between the public and private sectors. Most often these concerns are associated specifically with IPR and/or biotechnology, but similar considerations would apply to MTAs. They are all different aspects of the same set of economic forces.

Few actual surveys have been conducted to collect the experiences with information sharing under IPR. One, done about 1980, at the request of the US Department of Agriculture, focused on the effects of the US version of PBR (Butler and Marion, 1985, Chap. 3). This survey of public and private sector plant breeders concluded: '[PBR] appears to have had either a neutral or positive effect on the flow of information and germplasm from universities to other public and private plant breeding organizations ... [but it] appears to have prevented an increase in the flow of scientific knowledge from and materials from companies to universities, not a decline in absolute amount'. When delays occur, they are typically in the order of 30–60 days (Blumenthal *et al.*, 1996).

Further surveys are acutely needed in this area; however, hypotheses have emerged to explain why the flow of information is not as impeded as originally feared. First, the mutual dependence between the public and private sectors creates a long-term merging of interest. Second, the bulk of material is considered to have limited potential value. It has enough value to warrant some claim of benefit sharing through MTAs, but not enough to make the conditions onerous. And finally, university technology managers work diligently with their corporate clients to maintain freer access. Many universities (like the author's) have policies mandating the eventual public distribution of research findings, making private research a non-negotiable request. Firms funding the research may, however, be allowed a 30-day review prior to public release. In these and other ways, accommodations are being sought between commerce and knowledge generation.

Other expressed concerns focus on the relationship between IPR and biotechnology. Those concerns relate to (see, for example, Busch *et al.*, 1991, Chap. 7; Blumenthal *et al.*, 1996):

- shift in research emphasis from basic to applied;
- university/industry relationships;
- morality and ethics; and
- accountability.

Many of these issues go well beyond our consideration here and are thus not addressed (but see Chapter 9, pp. 170–173). It is, however, apparent that the broad roles and functions of MTAs are of widespread concern and will continue to be scrutinized for the foreseeable future.

Complex material transfer agreements

Matters of access, use and sharing become more involved for materials considered to have significant commercial potential. Additional conditions of MTAs can include:

- Limitation to research in a particular subject area or areas.
- Requirement of notification prior to publicizing an invention using the material. (Many countries disallow as non-novel inventions those which have had any prior public reporting.)
- Restriction on the kinds of research agreements with other parties using the materials.
- Reasonable access to research area and provision of relevant research data.
- Non-distribution requirement for recipient, and the recipient's institution.

Enforcing this range of use restrictions is costly and reasonable only for a very select group of genetic materials. In such cases, these detailed agreements do limit the movement of materials and knowledge, and might become problematic if implemented on a broad scale.

Sometimes, MTAs are intended to serve purposes beyond protecting rights to the exchanged materials. Downes *et al.* (1993), for example, have prepared a model for biodiversity contracting which envisages three parties: the user, the collector, and those who may have contributed to the identification of a commercial product, typically a local/indigenous community. The model draft agreement has several objectives including: research and commercialization agreements (royalties are established at the outset); establishing the ownership of any resultant patents; capacity building (involving local scientists in research with the materials); and conservation (conducting of an environmental assessment). In practice, these diverse stipulations would likely be distributed across several agreements, not just one. It might be particularly difficult to establish the share for a local/indigenous community as it is not formally a party to the contract. In many respects this model contract is similar to a Code of Conduct for a collector (see, above, pp. 21–27).

A third category of agreements involves an ongoing, evolving relationship between suppliers and users. The best publicized of these is the landmark Merck/INBio agreement for the provision of samples for pharmaceutical potential testing (see Laird, 1993; Sittenfeld and Gamez, 1993). Merck is a leading pharmaceutical company; INBio (National Biodiversity Institute,

Costa Rica) is an NGO with an agreement with the Ministry of Natural Resources, Energy and Mines granting some rights and obligations to use national parks and conservation areas. The multi-faceted agreement with Merck included the aspects detailed in Box 2.2.

Box 2.2. Major terms of the Merck/INBio genetic prospecting agreement.

- Provision of approximately 2000 samples to Merck over a 2-year period.
  Initial payment by Merck of $1,135,000 to be used for conservation, equipment, training and taxonomy. Of that amount, $130,000–180,000 was in equipment required for collecting and processing samples.

- Royalty payment upon successful commercialization of a compound resulting from a sample. The actual amount was never made public but is believed to be between 1 and 5%.

- Carrying out taxonomic studies using Costa Rican para-taxologists.

Sources: Laird, 1993; Sittenfeld and Gamez, 1993.

This particular agreement is frequently referred to, and as frequently critiqued, due to the widespread publicity. Yet as the areas of exploration are devoid of local and indigenous peoples, the agreement cannot serve as a model for national benefit sharing.

INBio is known to be involved in a number of similar agreements with other companies, but the specifics have not been publicized. Entities, both public and private, known or believed to be involved in contracting for genetic resources are listed in Box 2.3. Two of the firms, Shaman Pharmaceuticals and Andes Pharmaceuticals, warrant additional attention.

Shaman Pharmaceuticals was established in 1989 expressly to use ethnobotanical knowledge for identifying potential drugs. Particularly, materials used by at least three communities in different geographical areas are screened (King, 1994). Shaman also utilizes a programme for returning benefits to the communities assisting in the identification of successful products. The responsible entity is the Healing Forest Conservancy, a non-profit organization. Until the first products are marketed (pharmaceutical product development typically is a 12-year process), some benefits (reportedly 20% of the exploration budget) have been distributed at the onset. These include runway improvements for medical evacuations, which were requested by the community itself.

Initially, Shaman created considerable excitement, and raised significant sums of money from venture capitalists and a stock offering. This support

Box 2.3. Partial list of entities involved in contracting for genetic resources.

Name	Country	Materials used
Eli Lilly	USA	Plants, algae
Glaxo Group	UK	Fungi, marine, microbes, plants
Merck & Co.	USA	Fungi, microbes, plants, marine
Bristol-Myers Squibb	USA	Fungi, microbes, plants, marine
Ciba	Switzerland	Microbes, marine, plants
Rhône-Poulenc	France	Plants, marine, microbes
National Cancer Institute	USA	Plants, microbes, insects
Shaman Pharmaceuticals	USA	Marine, fungi, plants
Smith-Kline Beecham	USA	Plants, microbes
Andes Pharmaceuticals	USA	Plants

Source: Reid *et al.*, 1993, Table I.1.

seemed justified as several products rapidly entered clinical trials. In 1994, however, Shaman's pharmaceutical partner, Eli Lilly, did not renew its research contract apparently due to concerns about Shaman's ability to produce products in a timely and efficient manner, leading to a sharp stock price fall and subsequent restructuring/retrenchment leading to a withdrawal from the collection and screening of natural products. Asebey and Kempenaar (1995) see this as evidence that the Shaman model is not useful for countries wishing to use the sale of genetic resources as a means of economic development.

Andes Pharmaceuticals was established in 1993 with the specific objective of increasing value added in the supplying countries (Asebey and Kempenaar, 1995; Finesilver, 1995). Under the Merck/INBio agreement, INBio collected the materials and prepared samples, leaving the major value added to be done in the USA. In other cases, preliminary screens[5] are conducted in the supplying country, but seldom anything beyond that, placing the supplying (largely developing) countries in the role of providing resources for yet another class of products (Andes Pharmaceuticals, 1995).

---

[5] Screens are tests for biological activity. Until the recent past, that involved a slow and expensive process of subjecting growing cultures to the materials and observing any destruction. More recently, the process has been automated by using probes containing the structure of the target organisms (bacteria, cancer cells) to see if the extracts are of a shape which will bind themselves. Screens are increasingly specific, with the most particular ones held by companies as valued proprietary secrets. See Principe (1988) and Weiss (1995).

The Andes paradigm for natural products pharmaceutical discovery envisions quite the opposite scenario. Andes, in collaboration with its source country joint venture partners, will base both its biodiversity prospecting and most of its subsequent pharmaceutical discovery activities in developing countries rich in biodiversity. Activity is to be focused on Latin America, with Colombia the initial site of screening activities.

The proposed Andes arrangements – the process remains very new – go well beyond MTAs into joint venture agreements. The intent is the establishment of a partnership leading eventually to a division of activities with the supplying company conducting the bulk of the research close to the sources of supply and the partner providing principally marketing services. Over time and in another context, this was the arrangement which evolved for start-up biotech firms. Countries, however, unlike companies are not required to submit business plans, meaning countries sometimes do not adequately consider their goals and cooperative advantage in valued added activities (see Simpson and Sedjo, 1992). In the interim, a true partnership will be required for the ongoing enhancement in trust of the supplying company such that it receives the training and proprietary screens necessary for making major contributions in value added (Weiss, 1995). The number of these opportunities will be limited to the major pharmaceutical companies and the partnership arrangements they can manage.

Negotiating contract terms

Lesser and Krattiger (1994a) observed that, once the small number of major firms is adequately supplied, attention must be directed to smaller, more specialized companies:

> Transferring [genetic materials] to smaller firms in the north will require a more substantial effort. Greater resources must be put into identifying the buyers and tailoring products to their more particular needs. More specific to the topics discussed here, sellers will be expected to invest more in research and development, and be more knowledgeable about their products.

In an era when 'standard terms' for genetic materials have yet to arise – there is considerable controversy over the terms of the Merck/INBio agreement even though the specifics are not actually known – this presents a challenge to suppliers of genetic materials. The first and perhaps most difficult matter is the identification of objectives/expectations of countries and communities for their genetic resources. Is the intent the development of a new domestic industry along the lines of Andes Pharmaceuticals? (see above, pp. 34–35). Is it local control over resources and/or knowledge? Or is it principally a revenue-producing activity? Each suggests a different approach. For example, the goal of establishing a national industry requires careful assessment of comparative advantage (see Simpson and Sedjo, 1992).

These are case by case issues which must be resolved at the national and/or local levels. Payment terms (the division of remuneration between initial, lump sum payments) and royalties none the less can be structured to have different expected effects on total payments and the allocation of risks between suppliers and users. Negotiators should be aware of the ramifications of each scenario.

Lump sum vs. royalty payments[6]

Delayed (royalty) payments are generally preferred by the contracting firm, the multinational pharmaceutical or agricultural firm, although that need not be the case. Delaying payments means the firm reduces interest expenses on monies paid out. With the typical pharmaceutical product taking 12 years to reach the market, the interest costs could be large indeed. As important, initial payments shift the risk of failure to the firm (see below).

From the perspective of the supplier, the likelihood of a sample leading to a marketable product (the 'hit rate') is in the order of one in 10,000 (Sittenfeld and Gamez, 1993), making the risks of no future payment large indeed. For that reason, suppliers prefer initial payments. Lesser and Krattiger (1994b) demonstrate (but do not prove for all cases) that emphasizing initial payments reduces total expected payments to the seller as firms require compensation for accepting the risks of project failure. There may, none the less, be reasons for supplying countries/communities to prefer prepayments, even recognizing the likelihood of lower overall payments.

Prepayments provide a close association between the commercial agreement and its funding, which, by connecting value with the genetic resource base[7], is an incentive to conservation and careful use. Countries and communities may also have an urgent need for immediate money, and poor access to capital and/or high interest payments. Major firms on the other hand are typically well funded so initial payments can be viewed as a lower interest rate loan from the user to the supplier. Such an arrangement may make excellent economic sense, while recognizing initial payments favour the current generation over future ones.

Apportioning of risk[8]

Contract payment terms are associated with (i) imperfect and *imbalanced information* between the parties and (ii) *risk*. Risk, in the case considered here, is the uncertainty (both in terms of probability and time) of producing a valuable commercial product. Both the supplier of genetic resources and the purchaser (the pharmaceutical company in our example) invest in the hopes of discovering another penicillin, but have a high probability (risk) of

---

[6] This material draws on Lesser and Krattiger (1994b).
[7] Simpson and Sedjo (1992), however, warn against a 'once-and-for-all payment' which provides no incentive for further conservation.
[8] This material draws particularly on Sheldon (1996).

losing their investment (only 1/10,000 samples on average leads to a viable product, see above). Attitudes to this risk can be quite different. It might be expected the pharmaceutical company is what is called, in non-technical terms, *risk neutral*, i.e. has a symmetrical response to earning or losing a sum of money. Nations and communities, for their part, may be *risk averse*, thereby responding much more strongly to the chance of losing a sum than to earning a like amount. It can be demonstrated mathematically that, for efficiency, maximum risk should be borne by the more risk neutral of the parties.

An example helps clarify why different parties respond differentially to risk. If I were relatively wealthy, then a coin toss for $100 would not concern me as much as if the $100 was the family food budget. Or, perhaps I am involved in a number of coin tosses. Then I would expect the long-term statistical chance of 50/50 of winning, whereas in just a few tosses there is a real possibility of bad bets leading to a substantial loss. It would make a difference if I were betting my personal money, or someone else's. In general I might be more cautious with my personal funds. Or, if I were betting on another person's behalf, I would be attentive to how regularly my luck was observed. Watched regularly, I would be observed in some losing tosses and perhaps reluctant to try others; whereas if I were accountable only periodically I could again count on statistical probabilities of winning and losing half the time.

Admittedly this example is a bit extreme, but it does have some parallels with the genetic resource negotiating case. In general, large companies are well financed so moderate losses, though a concern, are not devastating. They also have multiple projects underway in recognition of the likelihood that many will fail with a few succeeding (the so-called portfolio effect). Finally, the company funds are those of stockholders for which management is held accountable for profitability over time; rarely do stockholders follow the prospects of a single product. Managing risk is what corporations do and are thus often risk neutral. The situation with government/community representatives is typically quite different. Licensing agreements are infrequent and closely watched; if successful, the credit may be taken by higher level politicians while any blame almost assuredly is focused on the responsible employees. It is no surprise then that governments are so reluctant to negotiate agreements. Indeed, it is sometimes rumoured that INBio, an NGO, was established in part to bypass potentially paralytic risk aversion by the government.

In addition to risk, informational issues are of concern regarding two particular factors. One is the case (known technically as *adverse selection*) in which one party is better informed than the other. For example, the experienced pharmaceutical company is likely to know better than the supplying country the likelihood of finding a valuable commercial product. How can an equitable agreement be reached if one party has an informational advantage over the other? Second is the uncertainty of how trustworthy, how reliable, is the other party (a concept tagged with the ominous sounding term of *moral hazard*). In this latter case, both the supplier and the buyer may be wary of the other, but in general there is more publicly available information on the Mercks of this world than on the INBios.

One of the complexities of contract structuring is the zero sum nature: what one party gains the other loses. This means terms must be structured carefully to apportion risk according to relative risk averseness while providing some assurances against adverse selection and moral hazard. The structuring process can be lengthy and complex, but both parties benefit in the long term from a properly structured agreement which allows a transaction to proceed. This is the incentive underlying the prolonged contract structuring process.

For our purposes here, the following (seemingly reasonable) requirements have been identified for the providers and users of genetic resources:

National/local/indigenous communities supplying the genetic resources are likely to:

- want up-front payment rather than waiting years for the uncertainty of royalties (drug development is on average a 12-year process to market);
- be risk averse;
- wish to screen samples to increase value added over time through training and equipment transfer;
- identify a reputable partner (moral hazard);
- be the poorer informed (adverse selection); and
- wish to claim the bulk of profits after the buyer share and other costs are deducted (residual claimant).

The users of genetic resources are likely to:

- want to delay expenditures rather than increase the risk of loss by making large up-front payments;
- identify a reputable partner, especially if transferring proprietary screening equipment (moral hazard);
- be risk neutral;
- be better informed (adverse selection); and
- wish to claim the bulk of the profits (be the residual claimant).

Determining which party retains the bulk of any profits (the 'residual claimant') is clearly a major factor in any negotiations. Operationally, the residual claimant is the one who pays the royalty; anything beyond that amount is retained. Conceptually, the two negotiating parties could be equal partners, but as a practical matter, that would seldom occur. The supplying country would not wish to provide half of the estimated $250 million cost of bringing a drug to market. Nor would a division of profits be widely acceptable, as the computation of the buyer's profits is too difficult for the seller to verify. Hence, royalties are typically based on the more easily determined sales.

There are several ways the residual claimant may be established. The claimant, which can be a person or factor of production like land, may be the input with the fewest alternative uses (lowest demand elasticity): agricultural land (value varies with crop production and price), taxi cab licenses (medallions) where rationed, beef feeder calves (the feed grain can be used in multiple ways, but not so with the calves). Alternatively, the residual claimant is the party retaining ownership, the apartment landlord as opposed to the tenant. In both cases the opportunity to retain profits also implies the opportunity to take losses. Also, since we assume the genetic resource supplier is risk averse, he/she will not simultaneously be the residual claimant in any standard kind of negotiated contract. This is an important consideration for countries and communities; they cannot simultaneously be risk averse and reap all the profits.

Taken alone, the risk adverse nature of the seller and the risk neutrality of the buyer would lead to the preferred form of payment of the up-front, lump-sum type. This is fine for the seller, but it incurs a moral hazard for the buyer; will the supplier actually deliver the quantity and quality (e.g. care in taxonomy, in any screens performed) as promised? The buyer can avoid that uncertainty by making payments on a piece work basis: so much per sample delivered. If the firm does not know what quality standard is appropriate, say when screening is just begun on a decentralized basis, then a relative standard based on (say) the average of all suppliers might be used (known as a tournament).

Simultaneously, the seller has a moral hazard; how reliable and trustworthy is the buyer? Frequently, government personnel in developing countries are wary of multinational firms. How might the contract terms help overcome that schism? The upfront payment can be seen not only as a financial benefit for the country, but as a guarantee of performance as well. A firm, like Merck, that is willing to risk the $1 million is indicating a level of trustworthiness; a scam (e.g. no intention of paying promised royalties) could probably be arranged at a lower cost.

Regrettably, the appropriate contract stipulations identified to this point are counter to each other: the seller wants upfront payments while the buyer chooses a piece work standard. This impasse could perhaps be overcome by making the upfront payment in the form of training and equipment. For the seller, this would indicate commitment by the buyer, while the buyer would recognize the seller could not benefit unless there was a serious commitment to process and deliver samples.

Remaining to be considered is adverse selection; how does one party protect itself against a possibly more knowledgeable partner? Here, the principal concern would be on the part of the seller negotiating with the experienced multinational company. One approach is to emphasize royalties rather than lump sum payments. This means eventual payments are made on the basis of commercial successes (sales), not on some general expectation (about which the buyer may be more knowledgeable). Even so, with an imbalance of information it would be difficult to identify the equitable/appropriate distribution between the lump sum and level of royalty payment. A real solution would lie outside the contracting process and involve improved information. A reasonable negotiating request by the seller would be for documentation of the underlying assumptions on value used by the buyer.

Conclusions and Approaches

By many appearances, the 'common heritage' approach to the sharing of the world's genetic resources without compensation or even control by the sources is approaching an end. The underlying cause appears to be technology, including biotechnology, which enhances the commercial value of genetic resources. Many mechanisms underlie this increase in economic value, including the option to transfer genes directly among unrelated species; the speed-up of techniques for screening materials for possible medicinal effects; and a general increase in the rate of bringing materials to the marketplace. But whatever the mechanism, the significant point is the increased commercial value[9] due to the use of genetic resources.

Observation, buttressed by economic theory, suggests that property rights systems often develop when resources acquire commercial value, and so it is with genetic materials. First came property rights for finished materials (outputs) in the forms of seeds, medicines, and the like. In some cases this was accomplished by extending the scope of subject matter (patents for plants) and sometimes through new forms (Plant Breeders' Rights). While the geographic scope of protection is presently limited, this is changing under pledges made as part of the GATT Uruguay Round (see Chapter 9, pp. 175–177). The existence of these pledges has additionally advanced the call for means to claim benefits for genetic resources, which are the raw material for many of these new technologies and products. Suppliers understandably consider it unfair that their material is not accorded commercial value while the resultant products can be quite profitable.

An assessment of the same intellectual property rights mechanisms used to protect the *products* of biotechnology from unapproved use, however, indicates they are seldom applicable to the *inputs*. While a full discussion is lengthy, the conceptual explanation is simple: IPR, for technical

---

[9] The societal value can be argued at length, and indeed is important. The focus here, though, is on the market (cash) value and the forces that puts into play.

and practical reasons, requires a use to be identified, but that seldom applies to the bulk of genetic resources about which little is known. The same general conclusion can be reached on the applicability of non-traditional forms of IPR, including expressions of folklore and cultural property, which too have not advanced to a practical form of protection. Thus, for good or ill, most protection will be achieved through material transfer agreements (MTAs) (UNEP/CBD/SBSTTA/2/7, 1996, Par. 86).

MTAs in themselves can be quite complex. They require access legislation to prevent the taking of materials without any permission (see Chapter 3). They potentially reduce the sharing and the flow of knowledge. Although delayed/reduced exchange under MTAs has not been documented to any degree, it is a significant matter which should be monitored periodically. And MTAs are difficult to plan and negotiate. The planning complexities relate to decisions on objectives. Is the objective simply to maximize revenue? Or does the country or community seek to establish a value-added business? How realistic is that? What government agencies would be responsible for this new area of activity?

Negotiations, once the components are identified, require a knowledge of standard practices and familiarity with the interpretation of terminology. This, in turn, implies the need for capacity building for many countries and communities.

The problem of negotiating is pronounced if the negotiators are government employees, bureaucrats. Governments tend to be risk averse, and tend to provide non-symmetric rewards to employees. If an employee negotiates a favourable arrangement with a multinational, he or she will be recognized and praised, to a degree. If, however, the agreement is perceived as unfavourable, criticism can be intense. Within such a framework, there is little incentive to do anything, which describes much of what is occurring now – very little. This is not a hypothetical matter, but describes the difficulty countries are having advancing commercial agreements for genetic resources (Box 2.4).

Lesser and Krattiger (1994a) have proposed the creation of a new, decentralized entity for assisting countries, on request, in these areas. They refer to this as 'Facilitator'. Such an entity could have the following functions:

- broker an agreement between two parties, potentially a government and a corporation;
- assist a country, on request, in the drafting of access legislation;
- provide assistance in the development of 'business plans' for private or community entities initiating work with genetic resources and requiring external funding and/or joint ventures with established firms;
- develop a targeted training and internship programme on enhancing negotiating skills for the parties directly involved; and

Box 2.4. Establishing an agreement for harvesting the Caribbean sea whip.

Following approximately 10 years of research by the Scripps Institution of Oceanography of the University of California, the Caribbean sea whip *Pseudopterogorgia elisabethae* was, in 1985, determined to possess anti-inflammatory and analgesic properties. The work was carried out under a research agreement with the Government of the Bahamas Department of Fisheries which specified the rights of the government to involve Bahamian scientists in the research process, and mandated a sharing of information. No claim of ownership or residual rights was made. The University of California system subsequently patented the findings in the United States (patent Nos 4,754,104 and 4,849,410). The University of California licensing office subsequently licensed the rights to two firms. One licence specifies rights to pharmaceutical uses; the materials are nearing the stage of clinical trials, which lasts several years. The second is to a firm that developed methods for extracting the useful compounds from dried materials and subsequently received US Federal Drug Administration approval for cosmetic use. The Estee Lauder Co. has a 2-year exclusive licence and has marketed the materials in a new product, *Resilience.* In the future the product will be accessible to other firms. There has been considerable interest expressed for using the extracted Caribbean sea whip materials in a variety of applications.

Under US regulations, cosmetic ingredients are generally developed from natural ingredients, that is to say, not synthesized. To date, harvesting has been allowed limited extended permits from the Department of Fisheries. Approximately 15,000 pounds (dry weight) have been harvested and exported exclusively through Marsh Harbour Exporters and Importers Ltd. Yield is in the order of 5–8% of dry weight. At a reported price of $25 a pound to fishermen, the product is the best paying of all seafood items and hence is popular with fishermen. Marsh Harbour has trained the fishermen in harvest techniques at the 50–100 foot depth at which these organisms grow. Because this depth is beyond the range of most fishermen and sport divers, and because *P. elisabethae* is non-edible and not in demand by tourists, it is little known and used throughout the Caribbean.

This sea whip, a soft coral, is quite abundant throughout the Bahamas. It is also present in unknown (but more reduced) concentrations and potency around the Caribbean region. Initial harvesting tests have indicated that cutting the animal four or more inches from the base allows it to regenerate; the exact regeneration period required is not known, especially for large areas. Ecological damage will be minimized, it is believed, because *P. elisabethae* is one of several species of gorgonian which occupy the same ecological niche so that its reduction could be offset by the other species. Recognizing that harvesting could reach 4000–5000 pounds a year or more, a thorough population and ecological study would be required prior to initiating extensive, ongoing harvesting. Such a study is estimated to cost $100,000–$200,000.

Box 2.4.   Continued.

Issues

The Department of Fisheries has received repeated permit requests for ongoing harvesting, but has not felt either sufficiently knowledgeable nor statutorily authorized to make such a decision. It is unclear just where within the Government of the Bahamas such authority and information resides. Among the complicating factors are the absence of funds for the ecological study and an uncertainty about what terms and conditions would be appropriate. At the same time, the licensee firms are understandably reluctant to invest substantial sums in a product over which they have only limited control. As regards terms and conditions, two could be involved – a payment for materials and some royalty on products sold. The former would be shared in some manner among divers, exporters, and the government. The latter would presumably flow exclusively to the government unless the funds were directed to, say, conservation or to employment creation. The employment creation could be in regards to partial processing of the animal, as employment is a significant issue in the Bahamas, especially on the outer islands. Any royalty agreement would have to be negotiated with the University of California system as the research agreement under which the discovery was made did not claim any residual rights.

In a broader context, the government needs to consider appropriate access/research legislation in the case of future discoveries. A number of terrestrial and maritime research groups have been active in the Bahamas over several decades. Indigenous/local rights issues have not arisen because the gorgonian is little known by the Caribbean peoples.

- foster the participation of scientific experts in the development of a plan for sustainable use of genetic resources subject to commercial utilization.

Differences of opinion exist as to whether the Facilitator should merely take an advisory role in negotiations, or whether it should negotiate on behalf of the supplier. While the final decision will and must be made by the owner of the resource, it should be recognized that a direct negotiating role would change the relationship with the private sector as well. The Facilitator could not be presented as neutral, and the contract information it collected would be scrutinized by the private parties to prevent undue advantage. There are merits in both roles, but both cannot be served simultaneously.

At the same time, the Facilitator should initiate the development of the following in-house capacity:

- information related to bioprospecting and other agreements and make non-confidential portions available for general use; and

- technical/legal issues, including marketing, technology, foreign investment, IPR, and property rights systems of target countries.

The intention for the Facilitator would be to begin with public funding, but transfer quickly to predominant fee-based funding under the concept that the services are of commercial value and should be paid for by beneficiaries. (Public benefit components such as assistance in developing access legislation would have a continuing need for financial support from the donor community.) At present the Facilitator exists as a concept only, one which may or may not mature to a functioning body. But whether the identified services are provided through it or through a diverse system of other organizations, the current reality is that countries require training and impartial, expert advice on genetic resources access to become the significant source of national benefits which is so desired by many source countries. The recent UNCTAD Biotrade Initiative is undertaking a number of these activities, including annual market surveys of contractual agreements, capacity building in negotiations, and IPR, among other topics, and an assessment of proposed exchange and prior informed consent mechanisms (see www.biotrade.org).

Chapter 3

Using Access Legislation to Implement Sovereign Rights Under Article 15

The discussion in Chapter 2 concludes with the projection that countries will, at least in the short term, use MTAs for controlling the use of and payment for genetic resources. (Resources held *ex situ* represent a special case under the Convention; see Chapter 6. MTAs, however, apply only when an agreement is specifically established.) Additional legislation is required to prohibit individuals from using genetic resources without permission, without agreeing on an MTA or other requirements established by national legislation. That controlling/limiting legislation is known as access legislation, for which countries are put on notice for adoption:

> **Article 15(1)** Recognizing the sovereign rights of States over their natural resources, the authority to determine access to genetic resources rests with the national governments and is *subject to national legislation* (emphasis added).

As Glowka *et al.* (1994; see also Chapter 2) note, national legislation regulating access is required; the reference in the Convention to the national 'sovereign right' is, in itself, insufficient to prohibit unauthorized use. Only when such legislation is in place can governments correctly describe unauthorized use as 'gene robbery' (see Jayaraman, 1994); unauthorized use is illegal only when so characterized in national legislation. At the same time, those sovereign rights under the Convention are limited by a requirement mandating environmentally sound use and the non-imposition of restrictions to transfer[1]:

> **Article 15(2):** Each Contracting Party shall endeavor to create conditions to facilitate access to genetic resources for environmentally sound uses by other Contracting Parties and not to impose restrictions that run counter to the objectives of this Convention.

The latter terminology raises some questions about the appropriateness of the 1994 decision by the African Ministerial Conference on the Environment to place a '... temporary ban on the transfer of any living African biological resource from the region until further notice and at least until a majority of African States have ratified the Convention and put into place legislation for appropriate provisions' (UNEP, 1994; Recommendation 10.1). The duration of the ban and its effective implementation are unclear at this time. A proposed moratorium at COP3 for the discontinuation of bioprospecting until ownership and related issues of indigenous and traditional peoples did not pass.

The purpose of this chapter is the delineation of issues regarding the form and function of access legislation. This is drawn in part from materials prepared in response to a request from COP2, discussed at COP3 and continuing on to COP4, and likely, beyond. The initial subsection includes a general discussion of the issues, the subsequent subsection, examples of existing legislation with, in subsection D, recommendations for how to proceed. As of the time of writing, only two countries (the Philippines and Ecuador (under the Andean Pact)) had detailed access legislation in place, and neither was yet known to have facilitated access under the new law. Hence, practical experience with the operation of the laws is very limited.

Four major approaches, strategies, have been employed by countries to date when developing access legislation (UNEP/CBD/COP/3/20, 1996, Par. 10). These include:

- specific legislation focusing on access;
- legislation incorporating a broader set of objectives, such as the implementation of the Convention;
- modification of existing legislation; and
- legislation intended primarily for other purposes, but touching on access matters.

Here, the focus will be on the first, the design of specific access legislation. This is done in part for convenience, as it is simpler to discuss in the

---

[1] Other sections of Article 15 refer to the need to achieve mutually agreed terms (15(4)) and prior informed consent (15(5)). Those matters are addressed respectively in Chapters 4 and 5.

abstract a free-standing law than one embodied in other legislation. This approach, of course, does not preclude the use of the concepts presented in this chapter in modifying existing legislation or in incorporating other, broader objectives. At the same time, there are definite limitations to making legislation too complex, to including multiple objectives in a single law. Legislation which is specific to particular forms of genetic materials (e.g. marine organisms) complicates the access procedures for potential users and invariably will lead to cases where jurisdiction is unclear across laws. The modification of existing legislation can lead to a similar trap. The USA followed that approach when adopting biosafety legislation on the conceptually valid argument that there was nothing fundamentally new about genetic engineering that could not be managed under existing risk-based regulations (58 FR 5878). The result, none the less, was jurisdictional complexity (and presented the potential of gaps) which could have been avoided by a new law which assigned jurisdiction. Hence, specific and focused legislation is advocated here, while recognizing that national considerations may require other approaches are followed.

Components of Access Legislation

For the purposes in this chapter, seven aspects of access legislation are highlighted for the development of a complete access-regulating process (conservation mandates would also be included, but are excluded here as being beyond the scope of this work):

- scope;
- meaning of access;
- mutually agreed terms;
- prior informed consent;
- benefit sharing, including sharing of scientific results;
- responsible authority; and
- penalties/sanctions for violations.

Scope

Scope involves at least two major components: (i) a delineation, the definition/identification of what materials require permission for transport and use (organism scope); and (ii) identification of those materials over which the government claims sovereign rights (ownership scope). Government ownership claims are not fixed but rather evolve with changes in law and practice. Thus, scope is both critical and complex to define in an unambiguous manner.

One approach (often used in IPR law) is to make the organism scope all encompassing, with specific exclusions. There is no reason why national legislation could not refer to materials in addition to those which are incorporated in the CBD Article 2 definition of genetic materials as, 'any material of plant, animal, microbial or other origin containing functional units of hereditary'. Some national laws (see below) claim rights to derivatives as well, but that may exceed the intent of access legislation to prevent unauthorized use. If derivatives were to be included in benefit sharing, the details could be established in the individual MTAs and not in the law.

A second set of exclusions would be needed to accommodate materials to which the government could not or did not extend sovereign rights, due to conflicting ownership, practical considerations, or other matters. As an example, sovereign rights to resources in the 200 mile territorial waters might be controlled by the Law of the Sea. Other materials might be subject to pre-existing ownership claims, such as IPR (patents, PBR), or treated as personal property. Also, there is the matter of practicality. There is little benefit to be gained from establishing sovereign control over materials abundantly available and dispersed (e.g. African violets), but defining which materials they are is a complex matter. An approach might be to refer to indigenous materials, but what is/is not indigenous is not always readily known. Perhaps reference could be made to materials which had entered the public domain by means of being described and/or referred to by name in publications, data bases and catalogues. Using this approach, taking seeds of a commercial variety of beans sold in a local market would be permitted (if not otherwise prohibited or restricted), but not the seeds of a local landrace.

An equally difficult question is the treatment of materials on private land. Countries have different traditions regarding the ownership of materials which occur on (or below the surface of) privately held land; some, for example, exclude resource rights from land ownership (e.g. timber in Honduras until recently, oil in Colombia, minerals in Argentina and in many other former Spanish colonies). The same national differences can be expected to be present in access legislation. Definitionally, the inclusion of materials on private land is not straightforward. But where it is excluded, there is no obvious prohibition, for example, to removing genetic materials found on private land from the country without permission. This matter could be addressed by making it a violation to remove such materials without the permission of the landowner, but the procedures for acquiring that permission could be different than for materials found on public property.

Access

Access refers to the kinds of activities for which permission is required (UNEP/CBD/COP/3/20, 1996, Par. 45). Activities include acquisition and use of materials identified under the scope discussion above. Two categories could be identified:

1. Research use, defined as undertakings for knowledge generation and/or other predominately non-commercial purposes, although recognizing that commercial uses are not precluded and may be identified inadvertently, and for which the suppliers are implicitly being asked to make a contribution in kind to facilitate that knowledge generation.
2. Commercial use, undertaken explicitly with the objective of developing marketable products or services.

In some respects the use of the terms research and commercial use is misleading, because all experimental work, like plant breeding, can legitimately be considered 'research'. Here the terms apply to the intended outcome of the research process, not to the procedures used. Of course, the terminology does not preclude the collection of payments for products identified under research agreements: that would be defined by the Material Transfer Agreement established. But it does provide for distinct prior informed consent requirements noted below.

Mutually agreed terms

Mutually agreed terms can be signified as having been established through submitting to the national Competent Authority a copy of a Material Transfer Agreement signed by both parties or their representatives. In this way, the Competent Authority functions as a regulatory body certifying that the requirement has been satisfied, without necessarily becoming involved in the specifics of the agreement. Several categories of national parties can be established depending on whether the resources to be transferred occur on public (including national parks), traditional, or private lands. Identification of those lands would be contained in other laws or regulations.

Prior informed consent

In Chapter 5, a case is made for establishing different prior informed consent (PIC) standards for transferring materials intended principally for research, as opposed to commercial, purposes. Commercial PIC would largely involve disclosure of intent and identification of expected and possible commercial uses of the materials, along with any known problems or risks. The responsibility for compiling information necessary for negotiating an equitable contract would rest with the seller, as is standard in commercial transactions. Many suppliers would require capacity building and information to negotiate effectively.

Research agreements, it is argued, are in a different category because their suppliers are asked to provide materials even though doing so is not expected to benefit them directly. This places the researcher in a role of having an elevated responsibility to the supplier, which is reflected in a higher information level requirement. The researcher may be required to provide information on:

- purposes of the research and anticipated benefits/beneficiaries;
- possible commercial applications of the research;
- any risks to the suppliers; and
- alterative approaches;

among other bits of information.

Specific PIC requirements, including the form and language of the requests, would be included in the implementing regulations and verified by the national Competent Authority. One verification component is documentation of *informed* consent. In Chapter 5 it is proposed that a two-step procedure, sometimes used by the medical profession, be employed. Part one is the information required and part two includes a kind of quiz in which the respondents are asked to reiterate the provided information in their own words and explain the decision reached, even if seemingly irrational. *Consent* can be indicated by signing the materials or by video-taping the process, as appropriate.

Benefit sharing

Principal benefit sharing arrangements are established in the MTAs, which are to be negotiated by the parties owning the materials to be transferred. Competent authorities could assist the negotiating process by: (i) developing model agreements; and (ii) making available on a confidential basis the terms of past agreements. Access legislation, however, is not intended as the benefit sharing mechanism; that is done through the negotiation of Material Transfer Agreements. Access legislation rather establishes the requirement for the completion of the MTAs and establishes and certifies compliance with prior informed intent requirements.

Countries may also choose to establish some benefit sharing as mandatory for all agreements. Examples may be a specification that all samples of materials be deposited in a national collection, or a requirement that national scientists are involved in all research projects, and/or that the information generated under the agreement be shared with the country of origin in specified ways.

Responsible authority

Access legislation requires an oversight body with several responsibilities in three general categories; the establishment of detailed requirements under the legislation; processing of applications with certification of requirement completion, as documented by the granting of a licence/permit; and, on request, the formal identification/designation/interpretation of the application of those requirements to specific cases. These are quite broad mandates, but are seen as necessary to clarify the requisite generalities of access legislation in specific cases. Clarification of requirements, along with certification of their completion in specific cases, is key to the establishment of workable requirements from a business perspective. Indeed, many business people will not begin an open-ended process where, say, a number of government agencies are authorized to negate an agreement but none to certify its completion (the 'Gatekeeper;' see OECD, 1996). The Competent Authority would have that authority.

To be useful, the Authority must be able to respond to requests in a timely manner. Provisionally, a 20-working-day time period could be used, extendable for an additional 20 days on submission to the applicant of the extension's justification.

Due to the broadness of the mandate, careful consideration must be given to the composition of the Competent Authority. The body will require specific competence, such as in national ownership legislation, as well as likely having representation from affected groups.

Penalties/sanctions

Meaningful legislation must have specific penalties for violations which are severe enough to serve as deterrents but are neither out of proportion for crimes against property nor a deterrent for entering in a negotiation in that country altogether. Penalties could be along the following lines:

- violation of an MTA or the provision of false and deceptive information in establishing the MTA resulting in the revocation of the agreement and payment of damages/treble damages, return of all materials;
- provision of false or misleading information for the establishment of PIC resulting in the revocation of the agreement and payment of damages/treble damages, return of all materials.

Complaints can always be redressed through the court system, but many governments make suing a government agency a difficult, if not forbidden, process. Yet there are a number of decisions to be made by the Competent Authority, such as the identification of the owners or designation of the category of owner, which should be open to appeal. The legislation could allow for a dispute resolution procedure such as arbitration for settling those matters. Wetter and Priem (1991) describe several approaches to such 'alternative dispute resolution' procedures.

Overview of Existing Access Legislation

The purpose of this subsection is the review of existing access legislation on a component-by-component basis. Legislation may be passed at several, and multiple, levels, including national, sub-national, and local (incorporating indigenous communities). Here, emphasis is on national legislation as national governments are the focus of the CBD, but other efforts described here provide additional insights/alternatives. Laird (1996, pp. 19–20) notes in particular the agreement of the University of Strathclyde Institute of Drug Research, the Kuna peoples, and the Awa Federation.

As of late 1996, only two countries had broad access legislation in place, the Philippines and the Andean Pact States (Colombia, Peru, Ecuador, Bolivia, and Venezuela) of which to date only Ecuador had incorporated the regional legislation into national law. These laws are evaluated in the greatest detail, but the status of developments in other states are also identified. In many cases, when legislation is under development, there is a significant amount of change from early versions to the final text, so the current status should not be evaluated too closely.

It should be noted that the legislation considered here is that directed to, or involving, benefit sharing. Many countries have and continue to use research permits, but with no mention of obligation for benefit sharing. The steps involved in securing permission for research use can be quite detailed and lengthy; as an example, McKamey (1996) provides a step-by-step description of what is required in Tanzania.

Gambia

The Gambia's National Environment Management Act of 1994 (Law No. 13/94) does not constitute access legislation *per se*, but does empower the competent national authority to prohibit or restrict any trade or traffic in any component of biological diversity (Article 32.g) (abstracted in UNEP/CBD/COP/2/13, 1995, Par. 18). Article 15 authorizes the Council to 'make regulations and prescribe guidelines regarding access to the genetic resources of The Gambia, including:

(a)   measures regulating the export of germplasm,
(b)   measures for sharing the benefits [...]; and
(c)   fees to be paid for access to germplasm'.

Costa Rica

The December 1992 Wildlife Protection Law (abstracted in Laird, 1996, p. 10) declares all wild animals to be in the 'public domain' and a 'national patrimony' while wild plants are viewed as 'public interest'. Collectors from a Conservation Area must obtain a permit from the Ministry of Natural Resources. Costa Rica has also formed an agreement with the Instituto Nacional de Biodiversidad (INBio) for the utilization of materials in Conservation Areas (see Chapter 2).

Cameroon

Legislation was passed by the National Assembly in January 1994 (Law No. 94/01 on Forestry, Wildlife and Fisheries) with implementation legislation endorsed that August (Implementing Decree No. 94/436/PM) (abstracted in Laird, 1996, p. 9). These provisions were hurriedly appended to the Forestry Law following the discovery by the US National Cancer Institute of a potential anti-HIV compound (michellamine-B) in a forest vine. The administering authority and interaction with other laws are unclear. The law (Article 12) declares all genetic resources (undefined) to be property of the State. A permit is required for use with benefits shared with the Government. Legislation distinguishes between sample collection for research purposes and commercial exploitation of a genetic resource. Research permits (Article D.13) require, among other steps, deposit of a sample within Cameroon, certificate of origin and, in the case of positive results, a 'bioethnologic' survey must be undertaken.

Brazil

The Draft Bill of Law on Access to Brazilian Biodiversity (No. 306 of 1995; abstracted in UNEP/CBD/COP/3/20, 1996, Par. 16) includes requirements regarding Access to Genetic Resources (Chapter 3) which involve prior informed consent, including the consent of local communities. The Bill includes benefit-sharing requirements that involve the participation of the country in the economic, social and environmental advantages of the products and processes obtained through the use of Brazil's genetic resources.

Fiji

The Draft Sustainable Development Law (15 May 1996; abstracted in UNEP/CBD/COP/3/20, Par. 17) includes access components in Part XIX, biodiversity conservation, and national parks management. Outlined is a general process for obtaining a permit for biodiversity prospecting (Article 249), including public notification, prior informed consent, and export controls. Benefit sharing, in Article 249(1)(c), requires that 'a fair return is provided for any commercial exploitation of Fiji's biological resources'.

Australia

Australia operates on a federal governmental system, meaning that access and benefit sharing often require both national and state legislation. Presently, access is controlled at the state level while the export of genetic resources is under federal purview, specifically under the Wildlife Protection Act of 1982 (regulation of exports and imports) and the Customs Act (prohibited exports). However, many categories of genetic resources are not subject to export controls, including seeds, fruit, and certain materials derived from native plants. The Commonwealth State Working Group on Access to Biological Resources was established in May 1994 to identify benefits to Australians and develop appropriate control mechanisms (UNEP/CBD/COP/3/20, 1996, Par. 18–20). Studies by this Group were carried out as part of the draft National Strategy for the Conservation of Australia's Biological Diversity

At the state level, to date, Western Australia has passed the 1993 Conservation and Land Management Amendment Act empowering the Department of Conservation and Land Management to enter into exclusive agreements to commercialize flora, but not otherwise amending permitting processes. The Queensland government had been developing draft legislation but opted to await the outcome of the Working Group.

Australia is an example of a system with multiple controls applying to access and use of genetic resources (documents filed with the CBD Interim Secretariat). The Federal Government controls marine resources through the Guidelines and Procedures Relating to Foreign Research Vessels, except if in a marine park or marine nature preserve established under state legislation. Flora on Crown Land is the property of the Federal Government, but fauna belongs to the government only until it has been legally taken. In particular, a national Fisheries Act licence overrides other controls, although conflicting definitions of 'fish' and 'fauna' present potential interpretive problems.

United States

The USA has not ratified the CBD but, with a federal form of government, otherwise shares with Australia many of the complexities of managing the commercialization of genetic resources. To date, there is no national law controlling the use or import/export of genetic resources (beyond regulations intended to protect the environment and health). Indeed, given the governmental structure, the national government can control use principally on federal lands. Efforts are currently underway on one part of those lands, the national parks, and particularly Yellowstone Park.

Yellowstone contains a large number of thermal springs which have been the source of *Thermis aquaticus (Taq)*, used in *Taq* DNA polymerase, a means of rapid multiplication of DNA. Patented by Hoffmann-LaRoche as pLSG1, it generates an estimated $200 million in sales annually (Lindstrom, 1996). At least 10 additional significant commercial products are traceable to Yellowstone microbes. While not presently of commercial importance, Yellowstone Park is also a source of a new kingdom of microbes named *Korarchaeota*, the most primitive organisms discovered to date (Milstein, 1995a).

Also, to date, the National Park System has been unable to benefit financially from any valuable materials found in its areas. Indeed, the 1916 establishing legislation has been interpreted to prohibit the taking of plants, fish, wildlife, rocks or minerals from the national parks (CFR 36 2.5(a)). The single exception is the 'specimen collection permit' (CFR 36 2.5(a)). A permit may be issued to a 'reputable scientific or educational institution' provided the collection will not result in damage or the materials are readily available outside the parks (CFR 36 2.5(b)). 'Specimens and data derived [...] will be made available to the public ...' (CFR 36 2.5(g(2))).

'Reputable scientific or educational institution' has been interpreted to include private companies, yet the Park Service presently lacks the authority to require compensation from those companies in the case of inadvertent commercializable discoveries (Lindstrom, 1996):

> The National Park Service does not encourage commercial development of natural resources within its jurisdiction. If, however, during the course of investigation, researchers make a commercially significant discovery, a means of sanctioning that discovery is now available through their research permit agreement with the superintendent and according to ongoing revisions of the Code of Federal Regulations.

Reports that royalties have already been collected are thus premature (e.g. Milstein, 1995a). Firms have, however, proposed voluntary contributions, and that is considered as a possibly viable approach (Milstein, 1995b).

Those revisions, now under ongoing review, propose the addition of the following language:

> (d) Specimens will remain Federal property unless [...] ownership is conveyed in writing ...
> (e) Regardless of ownership, the National Park Service retains the right to royalties ...

There is no mention of exclusivity of access in either the current or proposed version of CFR 36. The revision, however, drops the 2.5(g(2)) requirement that specimens and related data be made available to the public.

If adopted, the new procedures would involve three instruments:

- collection permit;
- memorandum of understanding setting out the purposes and benefits of the agreement; and

- contract or MTA modelled after INBio's contract with universities.

The MTA suggests a 1% royalty on sales and is voluntary in the sense that no enforcement mechanisms are proposed. By mid-1997, Yellowstone Park had established a commercialization agreement with Diversa giving $175,000 over 5 years plus royalties up to 10%.

The Philippines

The Philippines was the first (on 18 May 1995) country to adopt specific access legislation in line with the CBD in the form of Presidential Executive Order No. 247, 'Prescribing Guidelines and Establishing Regulatory Framework for the Prospecting of Biological and Genetic Resources, Their By-Products and Derivatives, for Scientific and Commercial Purposes, and for Other Purposes' ('Philippine Regulations'). The use of a Decree rather than legislation is said to make the procedures more flexible, so that the process can be assumed to have a degree of mutability. Due in part to this early adoption date, these Regulations have received considerable attention. Implementing regulations, 'Implementing Rules and Regulations on the Prospecting of Biological and Genetic Resources' ('Philippine Rules'), were subsequently adopted in June 1996 (abstracted in UNEP/CBD/COP/3/20, 1996). To date, no transactions are known to have occurred under these regulations and rules.

Given the significance of this governmental action, it seems appropriate to consider the aspects in depth, according to the components of legislation identified in subsection A above.

Scope

Regulations. Section 1, 'biological and genetic resources'.

Rules. Section 2, including 'genetic resources, organisms, or parts thereof, populations or any other biotic component of ecosystems, with actual or potential use or value for humanity such as plants, seeds, tissues, and other propagation materials, animals, microorganisms, live or preserved, whether whole or in part thereof'.

By-products: 'any part taken from biological and genetic resources … including compounds indirectly produced in a biochemical process or cycle'.

Derivative: 'something extracted from biological and genetic resources … taken from or modified from a product'.

Section 3 (a), 'prospecting of all biological and genetic resources in public domain, including natural growths on private lands'.

(b) '[E]xcept traditional use, all bioprospecting activities aimed at discovering, exploring, or using these resources for pharmacological, development, agricultural, and commercial applications'.

Meaning of access

Regulations. Section 3, 'prospecting of biological and genetic resources shall be allowed when the person, entity or corporation ... has entered into a research agreement'.

Two forms of agreement are recognized, Académic Research Agreement, when 'intent is primarily for academic purposes'; limited to Philippine entities; and Commercial Research Agreement, applicable to all 'private persons and corporations, including all agreements with foreign or international entities'.

Mutually agreed terms

Regulations. Section 5(e), 'The agreement shall include a provision for the payment of royalties to the National Government, local or indigenous cultural community and individual person ... Where appropriate and applicable, other forms of compensation may be negotiated'.

Rules. Section 8.1.11, 'upon mutual consent among the Philippine government, communities concerned, and the principal'.

Prior informed consent

Regulations. Section 2, prospecting shall be allowed: (a) 'within the ancestral lands and domains of indigenous cultural communities only with the prior informed consent of such communities; obtained in accordance with the customary laws of the concerned community' and (b) 'allowed only with the prior informed consent of the concerned local communities'.

Rules. Section 2 includes private landowners with indigenous peoples and local communities from whom PIC is required. Section 2 defines Indigenous Cultural Communities or Indigenous Peoples as 'a homogenous society identified by self-ascription and ascription by others, who have continuously lived as a community on communally bounded and defined territory, sharing common bonds of language, customs, traditions, and other distinctive cultural traits, and who, through resistance to the political, social, and cultural inroads of colonization, became historically differentiated from the majority of Filipinos'. Section 5.2 makes it the responsibility of government agencies to 'see to it that the consent required is obtained in accordance with the customary traditions, practices and mores of the concerned communities'.

Benefit sharing, including sharing of scientific results

Regulations. Benefit sharing is included in some detail in Section 5, including:

(b)   A complete set of all specimens collected shall be deposited.

(c)   Access to collected specimens and relevant data shall be allowed to all Filipino citizens and the Philippine governmental entities.

(d)   The ... collector must inform the Philippine Government ... if a commercial product is derived from such activity.

(k)   If the Commercial Collector or its principal is a foreign person or entity ... scientists who are citizens of the Philippines must be actively involved.

(l)   In the case of endemic species ... technology must be made available to a designated Philippine institution and can be used commercially and locally without paying royalty ... other arrangements may be negotiated.

(o)   A fixed fee must be paid to the Philippine government.

Rules. Section 8 includes a number of components drawn from the CBD, including participation in research, sharing research results, deposit of specimens, support for research into conservation and sustainable use, strengthening institutional capacity, and the donation to national institutions of equipment used as part of research, among other requirements.

Responsible authority

Regulations. Section 6 calls for the establishment of an Inter-Agency Committee on Biological and Genetic Resources with representation from government ministries, scientists, NGOs, and a People's Organization with indigenous peoples' membership.

Section 7 establishes the Committee's powers and functions as processing applications, determining that Research Agreement conditions are met, verifying PIC requirements, involving local scientists, and overseeing administrative and planning tasks as needed.

Rules. Section 10 essentially repeats the duties listed above.

Penalties/sanctions for violations

Penalties are generally not specified except the Regulations (Section 5(f)) allow for the termination of an agreement 'on the basis of public interest and welfare'.

Andean Pact

Also known as the Cartagena Agreement, the Pact legislation represents the first effort to standardize access laws across a region. While binding immediately following publication in the *Official Gazette*, implementation requires adoption of similarly worded secondary legislation in national law, something only Ecuador has done to date. Decision 391, 'The Common System on Access to Genetic Resources' was approved in July 1996 following several years of discussions and major revisions.

Scope

Article 1 defines genetic resources as 'any biological material containing genetic information of actual or potential value' and country of origin as the country possessing the materials in *in situ* conditions. Scope, defined in Article 3, includes genetic resources and their derivatives and intangible components but only to those 'of which the Member States are countries of origin'. Article 4 explicitly excludes human genetic resources.

A derivative is defined as (Article 1) 'a molecule or combination or mixture of natural molecules, including raw extracts of living or dead organisms of biological origin, derived from the metabolism of living organisms', while a synthesized product goes by the description of 'a substance obtained by means of an artificial process, using genetic information or other biological molecules. This includes semi-processed extracts and substances obtained through treatment of a derivative using an artificial process'.

Meaning of access

Access (Article 1) is the 'acquisition and use of genetic resources' from both *in* and *ex situ* sources.

Mutually agreed terms

These are incorporated implicitly under benefit sharing.

Prior informed consent

Article 15 requires the applicant to provide 'correct, complete and trustworthy' information, with the detailed requirements specified in Article 26.

The basis for PIC for indigenous and local communities, and private landowners are provided in Article 41(a). Detailed procedures must be included in national laws, in accordance with national rights.

Benefit sharing, including sharing of scientific results

The creation of 'the conditions for fair and equitable sharing of the benefits accruing from access' is an objective of the legislation identified in Article 2. Actual benefit sharing is on a two-tiered level, governmental and private/communal. Two permits are in fact required for access: a national permit granted by the Competent Authority and a subsidiary authorization granted by the landowner, administrator of a traditional land, or owner of mobile germplasm, such as livestock (Articles 41–44), as permitted by national law (Article 7). Specific requirements for national benefit sharing, specified in Article 17, are similar to those identified above for the Philippines.

Responsible authority

Article 1 defines a Competent National Authority for each member state which has the authority and responsibility to sign access contracts and to ensure compliance with the system requirements, as well as overseeing the granting of PIC. Tasks are administration, negotiations and granting (Article 50). Directors of the Competent National Authorities constitute the Andean Committee on Genetic Resources with policy, technical and capacity building responsibilities (Article 51).

Penalties/sanctions for violations

Article 47 describes the possible sanctions as cancellation or revocation of the permit, fines for damages, as well as civil and criminal penalties, when applicable.

Assessment of Current Access Legislation

From the proceeding subsection it is evident that access legislation has not evolved very far to date, and that no existing law provides a clear model for other countries to follow. This latter point is regrettable as there are significant benefits to the international harmonization of access legislation, even recognizing the necessary national differences which must be incorporated. Harmonization will certainly facilitate access by users, reduce the learning needed for each successive country, and reduce possible competition among countries for minimizing requirements. Suppliers, too, could draw more readily on the experiences of other countries, and capacity building activities could be streamlined. Given this, it is helpful to examine the legislation abstracted above for the identification of strengths and weaknesses. That is done for the detailed Philippines and Andean Pact laws in the spirit that countries at a less advanced stage of legislation development can benefit from an understanding of what has preceded them.

Both laws have explicit mandates for a national competent authority with the responsibility to sanction the completion of the access requirements, among other tasks. None the less, the Philippines' mandate for PIC with indigenous groups 'in accordance with the customary laws of the concerned community' does raise questions about how the potential user is to identify those 'customary laws' and assure their satisfaction. The Competent Authority or the government agencies as specified in the implementation regulations should be involved in assisting would-be users in that regard. It will be important to follow how this requirement is implemented. If implemented unsuccessfully, as a practical matter, traditional lands will be shunned by users. The Philippines' regulations do distinguish between research and commercial use, but their conditions for research use are very limited.

Both the Philippines law and the Andean Pact regulations define product scope in detail. When major international genebanks are situated within countries with access legislation, as is the case with the Philippines and Colombia, the legislation should make reference to the access status of those collections.

Claims of rights to derivatives and synthesized products raise some interesting strategic questions. Conceptually, access legislation controls access to the genetic resources, the initial permitting of prospecting if you will. The materials so collected will evidently originate from living material, whether it be in the form of parts, extracts, etc. The Philippines' definition of 'genetic resources' and 'derivatives' would seem to encompass well those materials. However, the Philippines' use of by-products ('including compounds indirectly produced in a biochemical process or cycle') as well as the Andean Pact's use of the terms 'derivative' and 'synthesized products' go far toward treated, possibly commercial, products. It would seem clearer to regulate access through this legislation and determine through negotiation when royalties may be owing. In any event, it is not possible in legislation to define all forms for which benefit sharing may be required. The approaches used would seem to increase complexity with little benefit.

One of the more difficult aspects of this legislation is that of benefit sharing. Legislatively, benefit sharing is divisible between those mandated by the State (partially in kind) and payments to communities and individuals. Mandated benefits include capacity building, use of national scientists, donation of equipment, supply of samples, access to samples and related information, among others. Whatever the merit of those benefits, they do diminish residual benefits available for local/indigenous communities and individuals. Otherwise, both the Philippine and Andean Pact laws allow for royalty payments (with other forms of compensation, as appropriate) to the government and/or local and indigenous communities. The mechanism for royalty payments in the Andean legislation, the use of two agreements, is clear while that for the Philippines is unspecified. However, users in Andean Pact nations are faced with two sequential negotiations: first with local/indigenous communities or private landowners, and second with the government. This will involve some complex negotiating strategies given the incentive for both sellers to demand terms which, combined, may exceed the value of the materials for the buyer. The local/indigenous or private sellers who negotiate first have an advantage, but governments are unlikely to accept outcomes where they are residual claimants.

Consider, as an example, the following scenario where the buyer is willing to offer a maximum royalty of 5%. Otherwise there is no interest in proceeding. The indigenous community where the materials are found initially requests 5% royalty and, only after protracted negotiations, concedes to 3¼%. The community feels it has made a major concession by selling a third below expectations. The government, however, is limited to the residual of 1¾%, which may well be considered inadequate. What will happen? Clearly, negotiations will be complex with considerable posturing, exactly the wrong environment for inexperienced negotiators. Alternatively, legislation/regulations could allow for concurrent negotiators (possibly in adjoining rooms and utilizing a mediator), a prior negotiated split between the government and community/individual followed by a single negotiation with the buyer, or a legislatively mandated split (50/50 or some other sharing).

Secrecy and exclusivity of agreements can have substantial commercial importance to industry. Certainly, any firm would carefully consider making a major investment in a product if a competitor had similar access and could patent first. (Most patent systems operate on the 'first-to-file' system, see Chapter 9). Similarly, allowing unrestricted access to samples and information for citizens, and royalty-free national use, as with the Philippines' decree, further reduces the commercial value of Philippine genetic resources. The Andean Pact legislation does allow for confidentiality when justified (Article 21).

Conclusions and Approaches

Access legislation is fundamental to the use aspects of Article 15. It serves, on the one hand, to prohibit unauthorized use by requiring a permit for collection and related activities. On the other hand, it serves as the mechanism for implementing benefit sharing and prior informed consent requirements. As such, it must operate on multiple levels, national and local. All this says that access legislation is not simple, yet undue complexity should be avoided as it encumbers the process and raises costs. The CBD warns against that end by requiring States not to 'impose restrictions that run counter to the objectives of this Convention' (Article 15(2)).

Only two entities – the Philippines and the Andean Pact – have adopted thorough access legislation. (The Pact legislation requires enabling legislation in the member states of Bolivia, Peru, Ecuador, Colombia and Venezuela, of which only Ecuador has enacted.) Several other countries have drafts in place (Brazil, Fiji), while Australia has advanced the process but must contend with the additional complexities of a federal (divided national/state government authority) government system. Given this, it is no surprise that there are no obvious models for adoption, but they provide a good starting place for succeeding legislation.

Both the Philippine regulations and Andean legislation provide for a national body with diverse membership to administer the laws and grant permits. Business considers it essential to have one office with the power to administer (or deny) permits. The Philippine legislation is, none the less, burdened by a vague PIC requirement for using traditional law with indigenous groups, a sweeping definition of scope which leaves the status of *ex situ* materials within the country unclear, broad sharing requirements for collected materials and associated data which reduce the commercial value, and stated, but unspecified, benefit-sharing arrangements with local/indigenous communities. There is separate treatment for academic-type and commercial agreements, but as only Philippine institutions qualify for research access the effect is limited. Overall, the legislation raises some issues which must be resolved through experience. This in my judgement will diminish commercial interest in Philippine genetic resources for the present, but the law certainly does not seem to impose significant restrictions in the context of the CBD prohibitor.

Thus the Pact legislation would appear the better model of the two. Yet the Pact has formulated a complex document of 51 articles which, too, will require interpretation before its full significance is apparent. Particular strengths are the regional approach which reduces complexities for users and competition among the States, as well as the limitation of scope to endemic materials, and which clarifies the genebank matter and attempts at claiming foreign materials. Benefit sharing and, indirectly, prior informed consent (albeit the means are not specified), are handled through a two-stage approval system requiring agreements with the owners, be they individuals, traditional communities or whatever, and the government. Provisions for confidentiality are established when needed for commercial purposes. There is no distinction between commercial and research use, but this does not constitute a major limitation.

This legislation (more than that of the Philippines) extends its scope to 'derivatives' and 'synthesized products'. Few would disagree that benefits should be claimed for products developed from or even suggested by genetic materials. Certainly science is replete with examples of organic materials leading to major contributions, in which regard one can cite aspirin among other commonly used materials. But the question is where the specifics of the agreements ought to lie – in the access legislation or in the individual MTA. To my thinking, access legislation regulates access to the genetic resources under certain conditions, among them that there be a negotiated benefit-sharing agreement. That agreement is where the specific compensation ought to be included, not in the access legislation, as derivatives will not be discovered in the wild or in farmers' fields. Claiming derivatives under access legislation could, on the other hand, create complexities. For example, if an MTA did not include some classes of derivatives, would the users have violated the access legislation and be liable for penalties? This is a real matter, for multiple classes of derivatives exist – those developed with simple, standard manipulations and those merely suggested by the natural compound. Each level of derivative would involve a differing contribution by the natural product and by the scientist/researcher. There are differences which should be reflected in the benefit-sharing agreements, not mandated in legislation. Also, at some extreme, the contribution of natural materials would seem to be exhausted in favour of the creativity of the researcher. Did the creator of the apple variety whose falling, according to legend, led to Newton probing the laws of gravity, deserve any of the credit?

Perhaps then the major limitation of the Pact legislation is its length and complexity. Following is my effort to draft simpler language encompassing the major strengths.

Scope

All living materials, parts, components, and extracts thereof found within the boundaries of *xxxxx* [country], its territories and possessions, but excluding:

- human genetic resources[2];
- materials acquired prior to the CBD entering into force in December 1993 and held in *ex situ* conditions[3];
- materials for which the sovereign rights of the government are established by a convention other than the Convention on Biological Diversity;
- materials for which other forms of ownership have been established, such as by IPR;
- materials deemed to have entered the public domain as indicated by reference to description and/or use of a variety name in publications, catalogues and data bases, as described in the implementing regulations.

Access

Permission is required for access, exportation and use of genetic resources identified under scope. Two categories of use are identified, as follows:

- research use, defined as undertakings for knowledge generation and/or other predominately non-commercial purposes, although recognizing that commercial uses are not precluded and may be identified inadvertently, and for which the suppliers of genetic resources are implicitly being asked to make a contribution in kind to facilitate that knowledge generation; and
- commercial use, undertaken explicitly with the objective of developing marketable products or services.

Mutually agreed terms

Prospective users of controlled genetic resources must document the reaching of mutually agreed terms with the providers of those resources through filing with the Competent Authority a signed Material Transfer Agreement specifying its terms and conditions. The MTA must satisfy other applicable national laws and regulations, including legislation controlling

---

[2] This exclusion was reiterated in COP Decision II/11.
[3] Following Article 15(3).

investments and ownership. Parties to agreements for controlled materials are identified within three ownership or use of rights categories for land and aquatic and marine resources:

- public;
- traditional/tribal; and
- private.

Prior informed consent

Prior informed consent requirements are related to the intended use of the materials as primarily commercial or research.

Commercial use is determined by the project description and agreement represented in the signed Material Transfer Agreement. Prospective users must disclose the intended use of the products produced, and any known risks associated with the collection of the materials or production/use of the products. Sellers are responsible for negotiating an agreement.

Research use is that undertaken for the generation of knowledge for which the supplier(s) are not expected to benefit directly. Because the suppliers are being asked in effect to make a public donation, the researchers are considered to have a higher information obligation than for commercial agreements. In particular, researchers are expected to provide the following information:

- purposes of the research and anticipated benefits/beneficiaries;
- possible commercial applications of the research;
- any risks to the suppliers; and
- alternative approaches;

among other information required by the Competent Authority and listed in the Regulations to this legislation. Suppliers are responsible to establish in the MTA: (i) the permissible conditions of commercialization and (ii) any remuneration or other benefits owing upon commercialization.

Informed consent is certifiable by either: (i) the signing of the disclosure form in the case of commercial agreements or (ii) the signing or videotaping of the presentation of information and consent by suppliers in the case of research agreements. The decision of which medium to use, paper copy or video, is to be made by and announced by the Competent Authority based on the Authority's assessment of the situation, and particularly the literacy level in the suppliers' native language. The Authority may also require that the consent process further include a restatement by the suppliers in their own words of the information provided and explanation of the decision reached. The Authority shall then determine if the level of the suppliers' comprehension was sufficient to provide informed consent, and if not, shall specify what additional steps would be required.

Benefit sharing

Benefits are to be shared with the national government and, where the resources occur on communal, traditional or private areas, are the property of autonomous government institutions or when mobile, are owned by a party other than the government, with the owners of those materials or their agents. Benefits claimed by the government on behalf of the peoples of *xxxxx* [country] may include:

- samples of materials collected and related data and information, provided that the Competent Authority establishes procedures for user requests for confidentiality;
- royalty-free domestic use if mutually acceptable;
- involvement of *xxxxx* scientists and/or technicians in collection and taxonomy, including training as needed and appropriate;
- negotiation of additional benefits payable on successful commercialization including possibly financial returns or returns in kind such as capacity building, transfer of structures and/or equipment. In cases where a benefit sharing agreement is also negotiated with the owners of the materials, the share received by the government, as determined by an impartial comparison of the terms of the agreements, shall not exceed *xxx* per cent of the total benefits.

Non-governmental owners are authorized to negotiate independently a benefit sharing agreement with the principals (users) which may specify monetary and/or non-monetary compensation, prior to and/or on successful commercialization, provided no other provisions of this law are violated.

Responsible Authority

The national government shall establish a Competent Authority for the interpretation and administration of this access legislation. The Authority shall have the following duties and responsibilities, plus others identified and delegated by the Government:

- serve as the national contact point for accessing genetic resources;
- establish the requirements for determining which agreements are research and which commercial and, on request, certify specific requests as one of the two;
- process applications;
- identify and make available national legislation and/or procedures describing the ownership or use of rights of genetic resources occurring on public, private and traditional areas;
- on request, identify the legal owners of materials located in specified areas and identify the parties to be involved in the negotiations;

- establish PIC requirements;
- certify the fulfilment of the PIC requirements;
- review submitted MTAs and designate the transactions as research or commercial.

The Authority shall be provided with the manpower and financial resources as required to discharge these responsibilities according to the timetable established in the regulations to this law. The Authority may be granted the prerogative by the Government to assess an applicant fee sufficient to cover the operating expenses.

Penalties/sanctions

Entities which do not fulfil the terms of agreements established under this Act are liable for cancellation of the access permit, payment of damages, and possible civil and criminal penalties. All non-criminal penalties, on mutual agreement, may be brought to binding arbitration according to the procedures of *xxxxx*.

Entities which export and/or use genetic resources covered by this Act without complying with it are liable for civil and or criminal penalties including, but not limited to, fines as specified in *xxxxx*.

Enforcement

The legislative focus is all on internal matters with little apparent attention to enforcement, of which detection must be considered a major aspect. As is discussed in Chapter 6, tracking materials either through genetic finger-printing or tracing commercial use is prohibitively expensive in general, even when the target materials are known. When the prospecting is focused on wild materials, tracking and detection of a party determined to circumvent the national regulations could effectively be impossible at the national level. That then raises the role and responsibility of countries receiving the materials. One approach is to require the showing of valid export permits for genetic materials to be imported into a country. Precedents exist in the Basel Convention, the UNEP London Guidelines, and the UNESCO Convention on Cultural property (see UNEP/CBD/COP/2/13, 1995, Par. 87).

The Basel Convention (Art. VI) and London Guidelines (Art. 7.3) forbid the exporting of a chemical without the sanction of the importing country (in those instances through providing evidence of PIC). With genetic resources, the process could be streamlined by requiring the showing of a valid exporting license, something common for a wide range of products. Downstream, companies could be required to maintain records of the source of genetic materials, paralleling the requirements of the UNESCO Convention on the Means of Prohibiting and Preventing the Illicit Import, Export and Transfer of Ownership of Cultural Property (Article 10). With genetic resources the process will be more difficult because the materials will change form so that, even with extreme care, not all agreements will be enforceable. The CIMMYT MTA (see Chapter 6) goes further in stipulating that third party recipients accept the same conditions agreed to by first line recipients, which is likely to be impossible to enforce. But the process proposed will at minimum provide a paper trail.

Enabling language could be included unilaterally in national access legislation, or operate on a reciprocal basis. Or it could be considered as a protocol under the Convention. Suggested enabling language follows:

> Importation of genetic materials covered by the Act, their parts and components, must be accompanied by a certificate from the Competent Authority in the supplying country attesting the materials were acquired in accordance with national access laws and regulations. Subsequent (third party) possessors of genetic resources must supply a copy of the original certificate of compliance. If the materials to be imported are excluded from the exporting country's access legislation, or the country has no such requirements in place at the time of the transfer, the potential importer must provide a sworn statement to that effect, subject to criminal penalties for false or misleading information.

Chapter 4

Valuation and Equity Under Articles 8, 15, 16 and 19

The term 'valuation,' containing the word 'value', is a far more complex matter than it might appear on first consideration. In purely private transactions, sellers often substitute the word value for price ('the value of a ... is ...') or combine the two ('value priced at ...'). Definitionally, price is the more straightforward concept, the figure established for an exchange. Value connotes a more judgmental assessment, a *fair* price or *worth* to the possessor. An object's value then changes according to the perceptions of the individuals making the assessment. Equity, too, has a judgmental aspect, for there are many perspectives on what constitutes equity in any particular circumstance.

In another more theoretical dimension, economists have developed multiple ways to characterize value, paticularly for non-marketed items. A major distinction is between use and non-use values. Non-use values include the benefits received from the simple knowledge that something (a remote wilderness) exists or will be available for use by future generations. Use values include indirect use (pollination by bees) and the familiar direct use. Direct use itself may be consumptive (destructive) or non-consumptive, as with ecotourism. In this chapter, for reasons explained in more detail in the following section, emphasis is on value in direct use.

Combining the complexities of the concepts of value and equity with the uncertainties of what constitutes genetic resources leads to an intricate matter indeed. Yet achieving the objectives of the CBD requires just that be done, a mandate repeated in four CBD Articles:

- 'equitable sharing of the benefits arising from the utilization of such knowledge, innovations and practices' (8(j))
- 'sharing in a fair and equitable way the ... benefits arising from the commercial and other utilization of genetic resources' (15)

- 'mutually agreed terms' (15 and 16)
- 'under fair and most favourable terms' (16)
- 'priority access on a fair and equitable basis' (19).

The purpose of this chapter is the exploration of the issues of value and equity under the Convention. Value is approached first, from both a conceptual and empirical perspective. This is necessary because of the need to determine just what is being valued – whose value over what time period. As will become apparent, these and other definitional matters greatly affect the estimated 'value' of genetic resources. Saying estimates range widely suggests lack of precision. Value estimates, even differing ones, none the less advance the goal of sustainable use of genetic resources in several important ways:

- if one dimension of equity is a fair price, then a knowledge of the value assists in the recognition of equity.
- achieving 'mutually agreed terms' is far more difficult if the parties differ greatly on the initial assessment of the value of the genetic resources to be exchanged. Providing impartial estimates of value will help establish a common negotiating base.
- value provides an incentive for preservation; the estimated value then indicates the degree of conservation incentive which can be anticipated from the marketing of genetic resources (McNeely, 1988).
- valuation contributes to the optimal allocation of funding for enhancing *in situ* conservation.

Of course, no discussion of 'equity' or 'value' will suffice in reconciling all parties to the outcome of a transaction. The matter is simply too personal for that to be accomplishable. However, it is hoped that participants will gain insights into their own and others' expectations. They will not necessarily agree on the equity of an outcome, but at a minimum, may understand the bases for their differences. Understanding in turn contributes to accommodation.

Clearly, this all applies only to monetary values, which ignores many of the ways the world's peoples relate to genetic resources. In some societies, selected genetic resources have a spiritual dimension, which is to say are non-market valued, or the society may not understand the concept of pricing (and selling) what exists in their common domain, as we do not think in terms of selling the air around us. Other societies, and individuals, accord non-human life forms a moral right to exist (Macer, 1990). Valuation, of course, has no place in such a world view. But if resources are to be exchanged for compensation, something the peoples of the world have demanded for some time, then valuation is a key aspect of that process. Indeed, in practice, human life is routinely valued in damage awards and safety measures (see, for example, Viscusi *et al.*, 1992, Chapter 20).

Approaches to Valuation of Genetic Resources[1]

The complexity of valuing genetic resources lies in several dimensions. As regards the *characteristics* of the materials themselves:

- the quantity in existence has been estimated only within orders of magnitude (see Chapter 1). As scarcity is often associated with value, it is especially complex to value something the dimensions of which are not defined.
- the usefulness of individual materials is known only probabilistically. The probability of finding a 'hit' when randomly screening materials for pharmaceutical use is variously estimated between 1/6000 and 1/30,000 (references in Lesser and Krattiger, 1994b).
- much material is redundant, meaning multiple near duplicates exist, so that many products serve as substitutes for themselves. With the core collections of genebanks, for example, it is estimated that 10% of the materials contain 70% of total genetic variability of the species (IPGRI, 1996a, Par. 37).

In another dimension, genetic resources are best described as *public goods*. Public goods are defined as those which are not depleted if used by another person (non-rivalrous); national defence is a classic example of a public good. Public goods are also non-exclusionary; individuals in a country cannot selectively be excluded from the shield of national defence. Because market-determined value is typically premised on scarcity and exclusion, the concept has little meaning for genetic resources. A related concept, seen from another perspective, is the distinction between *public (social)* and *private* value. The two equate only if:

- there are no externalities (e.g. pollution);
- there is perfect information;
- there are no biases in the formation of expectations; and
- all future costs and benefits are discounted at the social rate (Perrings, 1995, p. 832).

These are generally stringent conditions, but seem particularly inapplicable to genetic resources. Hence, at a minimum, private market values will typically not equate with social value.

The matter of *discount rates* is of considerable significance in the final valuation. The discount rate can be thought of intuitively (if not quite precisely) as interest rates; anyone having purchased an object with time payments is familiar with the effect of interest on the final cost. With genetic resources the situation operates in reverse; as many payments and

---

[1] These first two subsections draw particularly on Perrings (1995) and Pearce and Moran (1994). The theoretical underpinnings are from a branch of economics called public finance.

other benefits will not be made until some future date or periods, they need to be standardized in present value (discounted) terms. The issue is how to establish the discount rate.

Typically, public rates are lower than private ones because the public sector has a longer term perspective and because the public captures more of the value than can be appropriated by private parties. This latter point is especially true for public goods. I, for example, may pay admission to a comedy show, part of which goes to the comedians. They, however, cannot stop me from relaying the jokes to my friends and they to theirs (albeit in not as skilful a form). The public benefits (hopefully), but not the comedians. A similar argument can be made to describe why a *risk discount* would be calculated differently for public and private beings. In this context, risk refers to the likelihood of a genetic resource use being commercially successful, not to the likelihood of a public health risk.

Among individuals, the poor typically have a higher discount rate than the financially better off for the simple reason that the poor, with an uncertain immediate future, cannot have as long-term a perspective. Hence the accepted discount rate has an ethical component: should it reflect the immediate needs of the poor or the longer term perspective of the rich? The answer has intergenerational equity ramifications as well because it determines in part current vs. future consumption, the legacy of future generations. Here the perspective is taken that a discount rate equal to the marginal productivity of a resource is ethically neutral (Norgaard and Howarth, 1991).

For these and other reasons, value estimates of genetic resources have been calculated in many different ways. The major categories of valuation are described in Table 4.1. Which categories are included in any one value estimate obviously has great effect on the final figure so that these distinctions should be considered carefully when comparing the empirical estimates in the following section.

Recently, Simpson and Sedjo (1996) have innovatively proposed using search theory for valuing a marginal (e.g. 'the next') accession in plant breeding research. Search theory involves a balance between the costs of search (in plant breeding the cost of evaluating yet another accession for the presence of a desirble trait) vs. the expected benefit (say disease resistance). The benefit can be quite immense – Bagnara *et al.* (1996, Table 8) place the annual value of *all* international germplasm to the Italian durum wheat sector at US$476,000,000. However, Simpson and Sedjo (1996) are pessimistic regarding the empirical applicability of the search methodology due to data limitations and other factors. A principal tentative conclusion parallels earlier work that the value of an individual accession is likely to be small when the number of accessions is large (Simpson *et al.*, 1994). As further indications of possible future directions in valuation research, Brush and Meng (1996) propose surveying traditional farmers to ascertain

Table 4.1. Categories for valuing genetic resources.

Name	Description	Example
Use values		
Direct value	Value in direct use	Agricultural plants
Indirect use value	Value of supportive function	Pollination by bees
Consumptive use	Destructive use	Logging
Non-consumptive use	Non-destructive use	Recreation
Non-use values		
Bequest value	Benefit of anticipating future use	Parks, preserves
Existence value	Benefit of knowledge of existence, even if never used	Remote wilderness
(Quasi-)Option value	Preserve, if value uncertain	Biodiversity
Selected non-use measurement methodologies		
Revealed preference	Travel cost	Expenditures to recreate
Contingent valuation	Inquire of willingness to pay	Preserve tropical forests
Hedonic pricing	Unravel value from joint goods	Marginal value of air quality in different housing locations

their private valuation of landraces. That value, they propose, would provide a lower bound on the social research value. Smale (1996) further notes that a better knowledge of field level cultivar diversity is a prerequisite to economic valuation. Application of these approaches is some distance away, at best.

A final measure should be mentioned, the Total Economic Value (TEV), the summation of use and non-use values. When aggregating TEV, a decision must be made on the unit of measurement. Should it be per species, per hectare (as a unit of parkland), or all biodiversity? The more sweeping the estimate the more complex/less precise it is.

In terms of valuation, non-use values are the more complex to estimate. Values cannot be observed as in a market, but must be inferred indirectly from related behaviour. One of the earliest approaches was the travel cost method. What is a scenic area, or a public fishing access, worth to users? At minimum, it is what users spend to get there. This method has yielded useful estimates, but can lead to anomalies as well. Consider, for example, the case where the river lying near a city has better fishing than the more distant one; travel cost estimates will reveal that users are willing to pay more for fewer fish! While this is not a frequent outcome, it does indicate the importance of considering a number of studies rather than one.

Contingent valuation has become popular with the recent growth in experimental economics. For a long time, economists believed in the dogma that 'talk was cheap'; consumers' responses were not reliable outside an actual market situation. Markets, however, do not exist in all cases – for biodiversity in particular – so that alternatives are essential. In the USA, for example, recent law mandates a cost/benefit justification for environmental projects. When the benefits are extra-market, contingent valuation has been used for an approximation. Methodologically and theoretically complex, the concept of contingent valuation is straightforward; simply ask users their opinions (a good reference with example studies is Mitchell and Carson, 1990). 'What would you be willing to pay for ...?' Conceptually, almost anything can be given a value using this methodology. Other studies have used real money in exchanges. These would seem to be more accurate, including but not limited to tradeable rights, such as hunting and fishing licenses (see, for example, Heberlein and Bishop, 1986).

Problems have arisen, however, over the phrasing of the questions; should individuals be asked willingness to pay (WTP) or required compensation to give up what presently exists (WTA). 'How much would I be willing to pay annually to preserve the Amazon as it is presently?' vs. 'How much would I have to be paid to feel no loss at the destruction of the Amazon?' Common sense suggests there should be a strong symmetry between the two answers. Experience has shown otherwise; the figures can diverge by 10 times and more (Pearce and Moran, 1994, Table 3.2) as many of us seem to value items more once in our possession. This phenomenon can be explained by an inertia or familiarity factor, but it is troubling when the range of values can be so large based simply on how a question is posed.

A second problem is the abstractness of many of the questions. How does one place a reasonable value on an item or experience way outside one's personal frame of reference? What am I willing to pay to protect a species, the existence of which may well not be known to me? Chase *et al.* (1996, 1997) were able to overcome some of these difficulties through a careful selection of respondents. In a study of the value of Costa Rica's parks, rather than asking random individuals, they interviewed ecotourists in Costa Rica about the relative merits/value of the three most popular parks, with comparisons to a group of less and least popular parks (as measured by attendance). The result was a proposal for a graduated entrance fee system which attempts not only to balance visitations, but also to maximize the number of visitor days in the country, the real variable of interest to the government of a country where, since 1993, tourism has been the major foreign exchange earner.

For purposes here, emphasis is on direct use value for the simple reason that it defines most closely *willingness to pay*. Particularly, owners of genetic resources seeking compensation for sharing their materials would be principally interested in actual payments, not abstract concepts of value.

That, in most instances, is use value, the willingness to pay. This is not the only figure of consequence, but it is the focus of the negotiations resulting from access legislation (see Chapters 3 and 6).

Estimates of Value of Genetic Resources

The multiple available estimates on the value of genetic resources can be presented in several ways. The approach adopted here is first to report direct use values by type of use (pharmaceutical, agriculture, timber, etc.), followed by various measures of indirect use.

Direct use estimates are shown in Table 4.2.

These figures present only the direct values, which represent but a fraction of the estimated total values. The estimated total values, for a range of countries and habitats, are presented in Table 4.3.

Table 4.2. Estimates of direct use values of genetic resources by use category.

Category	Value per ...	Source
Pharmaceuticals	$770 extract (NPV, private, pre-tax)	Artuso, 1994
	$11B annually, plant-based medicines	Principe, 1991*
	$44, each untested species	Aylward, 1993
	$67–10,000, marginal species	Simpson *et al.*, 1994
	$20 maximum per hectare preserved	Simpson *et al.*, 1994
	$50–200/sample	Laird, 1993
	$7 ha^{-1}, minimum	Ruitenbeek, 1989*
	$1250 ha^{-1}, NPV, local, Belize	Balick and Mendelsohm, 1992*
Tropical forest	$6820 ha^{-1}, NPV, Peru sustainable harvest	Peters *et al.*, 1989
	$2455 ha^{-1}, Malaysia	Watson, 1988*
	$2306 ha^{-1}, NPV, Ecuador	Grimes *et al.*, 1993*
	$97 ha^{-1}, Brazil nuts	Mori, 1992**
	$44–117 ha^{-1}, India, gross value	Chopra, 1993**
Agriculture	$1M per accession (NPV @ 10%)	Evenson, 1994
	$67,000 per landrace	
Tourism	$1250 ha^{-1}, market value, Costa Rica	Tobias and Mendelsohm, 1991*
	$25M yr^{-1}, wildlife in Kenya	Brown and Henry, 1989*
	$10.325M yr^{-1}, wildlife in Zimbabwe	Child, 1984,* 1989*
	$8295 ha^{-1}, Virgin Islands park	Posner *et al.*, 1981*
	$42.7M w/fishing, NPV, Palawan, Phil.	Hodgson and Dixon, 1988**

Values in US dollars (unless otherwise indicated).
NPV – net present value.
* Quoted in Pearce and Moran, 1994; references therein.
** Quoted in Perrings, 1995; references therein.

Table 4.3. Estimates of indirect use values of genetic resources and habitats.

Value	Habitat	Included in valuation	Reference
$68–600B	Plant-based pharmaceuticals – worldwide	Economic value	Principe, 1991
$30B	Brazilian Amazon	CV/WTP	Gutierrez and Pearce, 1992*
$2.16M NPV/20 years	Hypothetical park (sighting lemurs)	CV/WTP	Kramer *et al.*, 1993*
£70 ha^{-1} yr^{-1}	Coastal marsh UK	Revealed preference	Danielson and Leitch, 1986*
Max $4.7M yr^{-1}	Thai elephant present	CV/WTP	Dobias, 1988*
$120 ha^{-1} yr^{-1}	Galapagos	Option value	de Groot, 1992*
$27 person^{-1} yr^{-1}	Air quality, Grand Canyon NPV	CV/WTP	Schultze *et al.*, 1987*
$2.8B	No oil spills	CV/WTP	Carson *et al.*, 1992*
$1300 ha^{-1} yr^{-1}	Forest carbon sink, Brazil	Indirect use	Pearce, 1992**
$13.9–19.2M	Protect marine tourism, Philippines	Indirect use	Hodgson and Dixon, 1988**
$148 person^{-1} yr^{-1}	Preserve old forests, spotted owls	Existence value	Brown *et al.*, 1994**
$9.3–21.2 person^{-1} yr^{-1}	Colorado wilderness	Existence value	Perrings, 1995, Table 12.5–4
$18.5 person^{-1} yr^{-1}	Preserve grizzly bears	Existence value	Perrings, 1995, Table 12.5–4

Values in US dollars.
NPV – net present value.
CV – contingent valuation.
WTP – willingness to pay (CV).
* Quoted in Pearce and Moran, 1994; references therein.
** Quoted in Perrings, 1995; references therein.

Not included in the preceding value estimates are assessments of the *total economic values*, the total of all use and non-use values. These figures can be quite substantial; Pearce and Moran (1994, Table 9.1) place an aggregate value on all tropical forests (net present value (NPV) over 20 years discounted at 5%) of about $13,000 ha^{-1}. That is significantly above the estimated NPV of $6800 for forest clearing and conversion to cattle ranching or related uses, suggesting preservation is the better economic decision.

Peters *et al.* (1989) estimated the net present value of a hectare of Amazonian rainforest used sustainably to produce fruits and latex at $6333, which is in excess of the sustainable timber production. Aggregation over broader areas, however, raises the issue of the price impact of supply, especially on the local fruit market. McNeely (1988) argues that 'perverse policies', those which encourage destructive use through favourable tax and other policies, are often responsible for countries making use decisions which are against their own long-term economic benefit. That is undoubtedly true in many instances, but it ignores the fact that non-use values include evaluations which will never result in actual payments. It may be a comfort to the inhabitants of tropical forests that residents of developed countries place a high value on the preservation of those forests, but that does nothing to provide income for daily life. On the other hand, those same developed country residents do pay cash for tropical hardwoods in the form of lumber. Another misperception exists, that the price paid by a private party should 'reflect the full value placed on it by society' (UNEP/CBD/SBSTTA/2/13, 1996). That would occur only if the private and social values are equal, a rare occurrence for reasons discussed above.

The great gap between use and non-use values can be gauged by comparing Tables 4.2 and 4.3. Whereas non-use values up to $30 billion are quoted, most use values are several thousand per hectare, or hundreds per biological sample. To these there are several outlier values which need further consideration: Aylward (1993) estimates a value of $23 million for untested species for pharmaceutical use, while Simpson *et al.* (1994) set the figure at $44. There are a number of technical assumptions about product possibilities and values which lead to these differences, but clearly the preponderance of the studies points to a far lower value. Indeed, even Laird's (1993) estimates of prices paid for samples – $50–200 each – may overstate current selling prices. This returns us to the preceding discussion of price vs. value, but certainly the willingness to pay is limited. Questions have been raised about the validity of any and all of these values, even Aylward's relatively high figure of $23 million. Applying Aylward's methodology to Ecuador, yields a value of $256 million. But with a few plausible changes in assumptions, that can be pushed to $429 million.

One other estimate is quite high, that of Evenson's (1994) for agricultural genetic resources of $1 million per accession. A unique characteristic of agricultural genetic resources pushes this value up so high. Due to the cumulative (add-on) nature of plant breeding, once a useful trait is incorporated, its contribution will continue indefinitely over huge areas. The first generation of IRRI rice varieties, for example, was planted on 65% of the Philippines' rice area, and by 1990, nearly 70% of rice varieties in widespread use had IRRI parentage (Evenson, 1994). Landraces, on the other hand, often contributed traits usable over limited areas, hence leaving a lower estimated value of $67,000 per landrace accession.

Those are the public or social values, and it has been the practice since the establishment of the CGIAR in 1960 to donate those varieties for use in developing countries (for a description of the seed development systems in developing countries see, for example, Cromwell *et al.*, 1992). The policy in developed countries has long permitted private seed sales, but value has proved difficult to capture for open pollinated crops, leading to the passage of Plant Breeders' Rights legislation (see Chapter 9). (Hybrids, which cannot be reproduced by farmers, have traditionally been far more profitable for seed companies.) The result of all of this is that in developed countries the willingness to pay for seeds has been significantly below their social value. Indeed, by one estimate, farmers are willing to pay only 25–50 cents for each \$1 of genetic improvement, thereby establishing a large risk margin for themselves (Hardy, quoted in Butler and Marion, 1985). Thus, even when the estimated public value of genetic resources is high, the willingness to pay can still be quite low.

In conclusion, there are a number of types of genetic resources values and measurement procedures. For self-funding of conservation – and as the basis of commercial transaction – the significant estimates are for the *private direct use value* of *individual sample* accessions. These values – referred to here as willingness-to-pay – are typically relatively low, often up to a few hundred dollars. Much work and methodological improvement are needed to improve the available estimates, but to date there are few indications the eventual values will rise significantly. The rather low market values suggest that (a) genetic prospecting should not be expected to provide major funding to support conservation and (b) transaction costs must be kept low so as not to absorb all the potential market value.

Other, and far larger, estimates apply variously to social values and to entire habitats. Those estimates are useful in a number of ways, including the allocation of public sector conservation expenditures. However, when considering commercial transactions, public and private use and non-use values should not be confused – they represent different perspectives.

Perspectives on What Constitutes 'Fair and Equitable'

The terms 'fair' and 'equitable' have culturally, even personally, based interpretations, making it difficult to define when that state has been reached, or even to recognize it after the fact. As a practical matter, this makes it tenuous to conclude any commercial agreements on sustainable use. It also makes it difficult to discuss the subject in a constructive manner, for there may be unstated differences in the underlying understanding of what constitutes 'fair and equitable'. Here four possible perspectives on fairness and equity are evaluated.

The four perspectives considered are:

1. Ability to pay;

2. Fair share;
3. No harm done; and
4. Marginal value.

This group is by no means exhaustive but does contain major approaches.

Equity and fairness must be evaluated in comparison to some norm, called 'welfare'. Welfare may be determined for groups, known as social welfare, or for individuals. A major issue arises in terms of how welfare is calculated. Conceptually, there are two major approaches. The first, which incorporates ability to pay and fair share (1 and 2 from the preceding list), is based on a judgement-derived social welfare function. Developing such a function requires value judgements, as can be explained in interpersonal terms. Those who believe a dollar is 'worth' more to the poor man than the rich are making interpersonal judgements. No matter how reasonable the expectation of the poor needing/enjoying money more than the rich may appear to be, there is no substantive evidence to support its universality; it is a value judgement.

The second approach, which incorporates our identified concepts of no harm done and marginal value (3 and 4 from the list above), is free of such value judgements regarding interpersonal value judgements. However, it does have the attribute of assuming welfare changes are independent of the underlying distribution of income. Under this approach, there are no rich and poor people; their financial status is irrelevant for welfare considerations.

Failure to recognize this basic distinction regarding interpersonal comparisons can lead to ongoing disagreements. Chapman (1994), for example, discusses the difficulties in regards to the International Covenant on Economic, Social, and Cultural Rights. It is fully possible, and indeed likely, that different national representatives were being completely sincere while discussing entirely distinct matters. This seems most likely when one group discusses equity from the perspective that interpersonal welfare comparisons cannot be made, while the other believes that it is the only appropriate approach.

While this discussion is presented in terms of income distribution, this should not be taken as an inference that income (money) is the only or even the most important equity issue. Financial compensation, either directly or as a medium of exchange, is, however, a convenient basis for discussing the concepts involved. And financial compensation, when appropriately structured, would satisfy the Convention's call for 'fair and equitable' sharing. Those interested in non-monetary compensation, or exchange in kind, can extrapolate from the approaches presented here.

With this background, the concepts and ramifications of interpersonal welfare comparisons are considered further.

With interpersonal comparisons

Ability to pay

Ability to pay concepts underlie so-called progressive tax systems under which higher income individuals pay in taxes a greater share of their income than lower income groups. The general concept seems to be that wealthier individuals can 'afford' higher payments, that there are certain essential expenditures and, with incomes above that level, individuals are indulging in less essential purchases. Progressive taxes then can be used to reallocate monies from higher to lower income groups. In countries without income taxes, the same effect can be achieved by taxing luxury products like cars, and subsidizing food for the poor.

Fair share

The notion of contributing one's 'fair share' has a strong fairness component. One contributes what one can. The complexity arises in deciding who makes the judgement of what is appropriate. Is it the individual – which can lead to broad disagreements – or is it in proportion to what one has, which reduces to the ability to pay? Moreover, the time period over which the contributions are made is an additional consideration. Should each event be independent, or is the share cumulative and long term? In particular, should the calculation transcend generations so that those, for example, who give and receive compensation are not necessarily the direct perpetrators and victims? Should the current generation of US citizens compensate the descendants of wronged slaves even though that worst of all human abuses ceased more than six generations ago? Clearly, there are many dimensions over which the 'fair share' can be determined, each based on some kind of value judgement, making broad agreement difficult to achieve.

From another approach, note that, as a general matter, residents of the ten poorest countries and their spokespersons – countries which garner but 2% of the world's GDP for their 5% of world population – understandably seek a larger income share. Conversely, inhabitants of the ten richest developed countries, which enjoy 60% of world's GDP for but 10% of population (1992 data, World Bank 1994, Tables 1 and 3), generally believe they have 'earned' the right to a higher standard of living. Both feel they are morally justified in their positions, meaning an agreement is elusive. For this reason too, it is difficult to develop broad support for value-based decisions.

No interpersonal comparisons

No harm done

One value-free perspective on equity is if, after the fact, all participants come out at least as well as they began, then there is not an injured or disadvantaged party. In economic theory terms, this is close to the concept of Parieto optimality which says, in its simplest terms, the only unambiguous way to advance group (social) welfare is for some individuals to be made better off and none worse off. Thus a technical innovation which netted me $3 billion but left the rest of the world's population unchanged would be Parieto optimal. So would an innovation which gave nearly everyone $1.00. Such a conclusion, it can be demonstrated theoretically, is equivalent to saying the underlying income distribution does not affect personal or social welfare.

An alternative, and slightly less restrictive approach, posits that welfare is enhanced if those benefited can compensate those who are disadvantaged. Examples are often given in monetary terms, although conceptually there is no reason to limit the benefits and compensation to money. A simple example would include a change benefiting 'A' by $100 but costing 'B' $50; since A can compensate B and still be better off, social welfare has improved. Two specific points should be made in this regard. First, achieving this kind of equity does not require the compensatory payments actually to be made. It is sufficient that enough benefits be generated to make possible the payments. Second, the assessment assumes income distribution does not matter, yet if A is wealthy and B poor, $50 may mean more to B than $100 does to A. Clearly on a governmental/societal level, there are severe limitations with this approach.

Marginal conditions

The marginality approach is really a matter of efficiency; if all economic systems operate flawlessly, then equity by this measure will be achieved. Efficiency requires efficiency in production so that nothing is wasted, efficiency in responding to users' requirements so there are no shortages and surpluses, and open exchange so those with a surplus of an item can exchange with those in a deficit position. Open exchange is not limited to a nation, but implies free trade as well. For the economic theorist, as this is a theoretical economic argument related to concepts of perfect competition, the equity argument can be extended further. Under competitive conditions, prices are equal to the 'worth' of goods, and workers are paid according to their contributions.

Begin, in a somewhat simplified explanation, with a production and cost function:

$Q = f(K, L)$, where Q = quantity produced, K = capital, L = labour;
$C = rK + wL$, where r is the unit cost of capital and w the wage rate, and C = cost.

Under perfect competition, by definition, long run profits are 0, and each firm has a minimum cost level of output, $\bar{Q}n$, so that the production process becomes a matter of cost minimization. Constructing the Lagrangian multiplier:

MIN $rK + wL - \lambda [(f(K, L) - \bar{Q}n)] = 0$, and differentiating with respect to K and L:

$df/dK = r - \lambda \, df/dK = 0$
$df/dL = w - \lambda \, df/dL = 0$, or
$r/w = df/dK/df/dL$, or $r \, df/dL = w \, df/dK$,

which says that the payment rate (r and w) is proportional to the contribution to output, or both capital and labour are paid their Value of Marginal Product (VMP).

The marginalist approach then characterizes a tidy world within which equity is formally defined. It has clear limitations as well, three in particular: (i) measurement in practice is difficult; (ii) the income distribution is assumed not to affect welfare; and (iii) there must be no uncompensated pollution, no non-priced components in the economy. Moreover, it assumes the economy is (perfectly) competitive; the conditions derived above do not apply under monopoly, for example (see any intermediate level microeconomics textbook). While these conditions can never be fully met, the approach does have the distinct benefit of suggesting equity may be enhanced by moving to a more efficient and open economic system. For the value-based systems discussed above, no such value-free statement about increasing social welfare could be made.

A key condition for marginality is the maintenance of competition. This is something for national governments to strive for in general. Specifically, as regards the Biodiversity Convention, effective competition between developing country sellers of genetic resources and developed country sellers of biotechnology products implies capacity building in negotiating and related activities for developing country representatives (see Chapter 2). Clearly, the marginality approach will not resolve the major ongoing issues as regards genetic resources, but it provides a clear, defensible basis on which to conduct individual transactions.

A competitive marketing system, as required by the marginality system, mandates a free (open) international trading system, as well as the internal conditions. That places an additional responsibility on national governments in association with the Uruguay Round of GATT (see Martin and Winters, 1995). Indeed, the Biodiversity Convention requirement for 'fair' as well as 'most favourable' terms (Article 16(2)) is linguistically close to the GATT requirement of *Most-Favoured-Nation* (MFN) *Treatment*. MFN Treatment specifies that 'any advantage, favour, privilege or immunity granted by any contracting party to ... any other country shall be accorded immediately and unconditionally to ... all other contracting parties'

(Article I(1)). Hence, internationally as well as nationally, there is a need for nondiscrimination[2].

Conclusions and Approaches

This chapter explores two intricate but key aspects of exchange under the Biodiversity Convention: What is the equitable sharing of benefits? What, indeed, are the benefits? These are frequent topics for conversation, but less easy to implement.

One place to begin, the one used here, is a consideration of the 'value' of genetic resources. An assessment of value quickly leads to multiple kinds of value, and estimation methodologies. Two of particular significance here are *use* (direct and indirect) and *non-use* (option, existence and bequest) values. Use values, especially direct use, are typically based on actual market transactions, while non-use values rely on other approaches, including asking individuals the value to them of an object or service. Total economic value is a summation of all the estimated component values. A second distinction can be drawn between *public* (social) and *private* value. Generally, individuals cannot capture all of the value of a product or service. A doctor, for example, by curing one sick child will help prevent others from catching the same disease. The doctor is paid for the single treatment, while society in general benefits from a healthier population. Public value hence typically exceeds private value.

A literature review of estimates of genetic resources values, either as individual items or in combination as an ecosystem, indicates the total economic value of a habitat can be very large. A figure of $30 billion has been proposed for the Brazilian Amazon. Much of this value, however, is of the non-use type. Direct use values are far smaller, sometimes a few hundred dollars a hectare. Smallest yet are direct use value estimates for individual samples/accessions, often also in the $200 range. Yet this final category is most relevant to commercial agreements and to self-funding of conservation.

While often small, the position taken here is that direct use values, referred to as *willingness to pay*, are more indicative of actual cash payments for genetic resources. Cash payments are not the only means of

---

[2] Some care must be taken interpreting these comments for experience – and theory of second best – show that a partial movement to competition does not necessarily improve, and may hinder, efficiency and equity. However, what is suggested here is not a full privatization of an economy of the scope often associated with structural adjustment, but rather a targeted deregulation and capacity enhancement in sectors associated with genetic resources. Such steps may involve an increase rather than a decrease in direct government involvement, at least initially. See related discussion in Rapley (1996), Chapter 4.

valuation, or of compensation, but they are important as a relatively easy measure for equity in transactions. Moreover, cash payments provide a clear incentive for the conservation of genetic resources. With relatively low estimated willingness to pay values, great care must be taken in the use of the available monies to maximize the incentive effect. Moreover, the low estimated value indicates that exchange transactions for genetic resources must be efficient so as not to absorb all the value of the materials themselves.

The other dimension of equitable benefit sharing is conceptualizing what the idea means in practice, in the context of a particular agreement. Several approaches are presented, differing on whether interpersonal welfare comparisons need be made or not. In general, systems using interpersonal comparisons are more personally satisfactory; each of us can rank others according to need or worthiness. Unfortunately, those rankings are so personal as to be difficult to achieve agreement across groups. Non-comparison systems do provide a more specific approach for securing equity, but with a notable limitation. The limitation is that wealth should not be considered in equity; the rich and poor should be treated as one. This says, in one context, that the poor villagers negotiating with the rich multinational firm cannot operate from the premise that the firm 'should' pay a higher price because the firm 'could afford to'. While that limitation is significant, it should be considered in the context of the benefits.

Here, the _marginal value_ approach to equity is endorsed. Considered conceptually through the strictures of economic theory, the marginal value approach promises that the contributors are rewarded according to the value of their contributions, certainly a plausible working definition of 'fair share'. The theory goes on to establish the requisite conditions. In addition to ignoring differences in wealth, and assuming that no unpaid components exist in the economy (meaning, among other things, there is no pollution), the marginal value approach requires the market to be perfectly competitive. Perfect competition is a theoretical concept which mandates, among other things, perfect knowledge and free mobility of resources.

Perfect knowledge, in the context of genetic resources, says that sellers and buyers must be informed of the characteristics of the resources, and the prevailing prices. This has not been true to date when the major multinational firm buyers are considered to be more knowledgeable of the market than sellers, creating an imbalance. The Clearing-house Mechanism mandated under the CBD (Article 18(13)) could in part assist in overcoming this informational imbalance. Otherwise, information on individual transactions, or even the existence of those transactions, will be difficult to acquire, as the negotiations are typically private. Indeed, the royalty rate for the Merck/INBio bioprospecting agreement, the most publicized of them all, has never been announced (see Chapter 2).

Governments will need to take additional steps to assist the openness of transactions for genetic resources transactions. A lack of openness will impede the equity of transactions. While admittedly difficult to implement, the enhancement of the openness of a market mechanism for genetic resources does provide governments with a procedure for improving the equity of any resulting agreements.

Chapter 5

Informed Consent Under Articles 15 and 19

Prior informed consent (PIC) is mentioned in but two subsections of the CBD, but the controversy surrounding this subject suggests a far greater significance. One recent (but non-plant) example is that of the Guaymi Indians of Panama. Blood samples taken in conjunction with the Human Genome Diversity Project, with the objective of collecting blood and hair samples of some 722 different peoples, 'isolates of historic interest', were subsequently subject to a patent application claiming a medical use. The application was withdrawn in 1993, and the samples subsequently returned, due to protests from the World Council of Indigenous Peoples and the Guaymi Congress (UNDP, 1994). Among the issues raised was the sufficiency of PIC for forewarning the donors of the possibility of commercial use. Nor need the donors be indigenous peoples. John Moore has become quite well known as the unwitting donor of tumorous spleen cells which were used as the basis of a successful and lucrative pharmaceutical product licensed to Sandoz. A court subsequently ruled the cell 'donor' was improperly denied PIC regarding the research interests in his cells, and was entitled to financial compensation (Burrows, 1997).

Use of plants and related materials does not excite quite that level of response, but the CBD's references to PIC none the less present some philosophical and practical issues which must be resolved to the satisfaction of a culturally diverse world. The references themselves are quite distinct, one referring to all genetic material and the second to the movement of genetically modified organisms (biosafety):

> **Article 15(5):** Access to genetic resources shall be subject to prior informed consent of the Contracting Party providing such resources, unless otherwise determined by that Party.

Article 19(3): The Parties shall consider the need for and modalities of a protocol setting out appropriate procedures, including, in particular, advanced informed agreement ...

One other convention, the Convention on the Control of Transboundary Movements of Hazardous Wastes and Their Disposal (Basel Convention), uses the term PIC (see also Chapter 1, p. 3)[1]. There, under Articles 6 and 7, a national authority must be identified for administering the procedures, and the information to be provided is specified in Annex V A. This is similar to what might develop under a biosafety protocol presently under consideration for Transboundary movements (UNEP/CBD/COP/1/4, 1994, Section 4.2.2). These are highly significant issues technically, but bounded within the context of regulations.

Far more open ended is the matter of PIC for access where the issues arise of who should be informed, how much information of what type is required, who should do the informing, what documentation is needed, and how can the supposition of informed consent be substantiated. Plus this must be accomplished without imposing '... restrictions that run counter to the objectives of this Convention' (Article 15(2)). Measures for answering those questions in a complete yet pragmatic way are the objective of this chapter. For insights we turn to the medical literature where informed consent has been both a philosophical and practical consideration traceable back to at least 1822. The approaches to be developed here are more applicable to the broader Article 15 issues as opposed to the regulatory-based Article 19 considerations.

Before proceeding, however, there is one matter not open to much debate. This is that the PIC requirement is optional ('unless otherwise determined by that Party'). Negating that requirement could be explicit – a statement of no PIC requirement – or implicit – no established requirement in the legal system (Glowka *et al.*, 1994, p. 81). Hence, many countries will wish to establish PIC requirements in their access legislation, as the Philippines has already done (see below, also Chapter 3). The following may assist in that process.

The material as presented here is somewhat narrower than what is sometimes inferred by the PIC requirement. PIC, for example, may be linked to indigenous/traditional resource rights – and hence to property rights – because to be consulted implies one has discretionary rights, e.g. control. However, in the context of the Convention, reference is to the consent of contracting parties (states) so that indigenous/traditional rights might better be treated where specifically identified, such as in Article 8(j).

---

[1] In addition, several other agreements and codes of conduct employ PIC, as follows. (See UNEP/CBD/COP/2/13, 1995, Par. 29): UNEP London Guidelines for the Exchange of Information on Chemicals in International Trade (Amended, 1992); IAEA Code of Practice on the International Transboundary Movement of Radioactive Waste (1990).

This is dealt with in Chapter 7. The discussion in this chapter is limited to a concept evaluation of the implications of PIC and suggestions for the implementation of specific practices.

Lessons From the Use of Prior Informed Consent in Medicine

Veatch (1976, p. 26-1) identifies three generally accepted theoretical foundations for informed consent. These refer specifically to medical treatment and research, but in a broader context apply as well to other interactions with individuals, such as the use of their sacred and traditional knowledge as well as other personal and community 'property':

- within the context of the Hippocratic oath[2] for medical practitioners, PIC can help in protecting patients from harm;
- an aspect of assuring the greatest good for the greatest number (utilitarian); and
- the individual's right to self determination.

The first justification, that based on protection from harm, could be more readily addressed in medicine by banning all non-therapeutic research; the primary purpose of which is not benefiting the patient being treated. There may be societal benefits from the research, just as there may be societal benefits from access to and use of genetic resources, but that is the rationale for the second justification, the greatest good. The general good argument follows the line that society will benefit from medical research, but that research is dependent in part on public support. Consent might help to insure that support, and should be employed.

There is a some logical inconsistency with that justification: if the underlying objective is the public good, that may better be served by *not* seeking consent. Conversely, the public good justification places no limits on the subjection of minority (including a minority of one) rights to the majority. Clearly public benefit is a societal consideration, but one not well addressed through informed consent. Hence, the overriding justification for informed consent is generally recognized to be the *right to self-determination*. Within medicine the focus is on the individual, but there is no loss in the principle by expanding it to communities like the Guaymi.

This justification leads to a *definition* (Reiss, 1976, pp. 25–27): '[Consent] is an affirmative agreement by free choice to provide information

---

[2] One version of the Hippocratic oath reads as follows: 'Under no circumstances is a doctor permitted to do anything that would weaken the physical or mental resistance of a human being . . .' (World Medical Association, 1949, quoted in Veatch, 1976, pp. 26–27). In Western medicine the oath operates something like a code of conduct in that it is not a legal standard but rather a guiding principle.

under stated or agreed upon conditions. For consent to be *informed* that means anyone consenting must be able to *predict* reasonably well ... what information will be sought and what risks or benefits will follow from participation, given only the information provided at the time consent is initially requested'. More formally, the US Government operates with the following definition (45 CFR 46.3):

> 'Informed consent' means the knowing consent of an individual or his legally authorized representative, so situated as to be able to exercise his free power of choice without undue inducement or any element of force, fraud, deceit, duress or any other form of constraint or coercion. The basic elements of information necessary to such consent are:
> (1) a fair explanation of the procedures to be followed, and their purposes, including identification of any procedures which are experimental;
> (2) a description of any attendant discomforts and risks reasonably to be expected;
> (3) a description of any benefits reasonably to be expected;
> (4) a disclosure of any appropriate alternative procedures that might be advantageous for the subject;
> (5) an offer to answer any inquiries concerning the procedures; and
> (6) an instruction that the person is free to withdraw his consent and to discontinue participation in the project at any time without prejudice to the subject.

These two definitions serve several critical functions. They help identify some of the kinds of information which need to be shared, and provide a test for the sufficiency of that information (prediction). And they present the information needs in a way that identifies PIC as a type of *contractual agreement*, or perhaps more accurately a negotiation process leading to a contractual agreement. But this treatment of PIC as a contract can be misleading in implying that PIC is a better established concept than it is, when in reality the notion and its implementation are ill defined (Levine, 1975, p. 7).

PIC further involves some anomalies when contrasted with the traditional contractual process. First, the party seeking consent (in Western medicine the researcher/physician; with genetic resources, a researcher or company) is also the best informed source of that information. Hence, the party is cast in an ambiguous dual role of both seeking an outcome and providing information suggesting why that outcome might not be in the consentee's best interest. Second, and related, typical negotiations proceed on the basis of the opposing parties taking personal responsibility for informing themselves and preparing a position and/or strategy. With PIC that direct responsibility is overridden in exchange for one of the contractors being held to a 'higher standard of responsible conduct' (Levine, 1975, pp. 3–7). The justification for this higher standard could be:

(a) the researcher/contractor is the sole source of pertinent information; or

(b)   the researcher/contractor is a readily available source of pertinent information; or

(c)   the researcher/contractor has a broad responsibility to the perspective consentee, a responsibility which requires that the consentee be helped in avoiding a potentially non-optimal decision.

The final justification (c) is in accordance with at least past Western concepts of the role of the physician as being in an elevated status compared with the patient. The physician was sometimes seen as having additional wisdom, both in technical matters and perspective, for guiding the patient to the best treatment. That paternalistic concept, however, is changing toward one of autonomous individuals who can make wise decisions given the proper information. The position of autonomous individuals also would seem to apply to suppliers of genetic resources. The standards thus must be based on the contractor as an information source (b). For a medical treatment, the physician may indeed be the only individual knowledgeable as to the exact procedures involved, and hence the risks/benefits involved for the individual (a). However, risk information is typically experimental so that any individual would have to rely on additional sources of information in addition to personal experience. In medicine, the scientific literature serves much of that function. That literature is publicly available, although some expertise would likely be required to interpret it. The contractor would have that expertise, but not as a sole source. Thus we have remaining the efficiency argument, the contractor as an efficient/convenient source of information (b).

The contractor additionally must anticipate what information the patient/ resource supplier should have to make an informed decision. This, too, raises multiple complexities, such as when information may operate against an optimal decision (see Levine, 1975). In the medical context, experiments may require that the patient be incompletely informed (i.e. that he/she is receiving a placebo). Or the patient may not request or choose to hear pertinent information; should it be provided anyway? Should a patient be informed in all cases that he/she is not expected to live more than a month, or given other information which might increase anxiety with no discernible benefit?

These questions cannot be answered in the abstract, and indeed need not be for the purposes here. The intent is rather to emphasize that PIC for genetic resource access is quite different from the traditional commercial transaction, which: (i) places the contractor in an ambiguous position *vis-à-vis* the supplier, and (ii) indicates the supplier is in some measures subordinate to and dependent on the contractor as an available source of pertinent information. Commercial transactions often incorporate means for equilibrating the information imbalance typically existing between negotiating parties (see Chapter 2, pp. 37–40). PIC for a research use is a different matter, at least to the extent the genetic resource suppliers are not benefited directly from the work. They may choose to make an in kind contribution to the work in the form of donated resources, but only under the specific conditions prescribed through PIC.

Can the Personal Autonomy Justification be Extended to a Non-Western Context[3]

The preceding discussion is based on Western concepts of the autonomy of the individual in society. May such a philosophical concept be applied to more community-oriented societies? Or must a different justification be found?

The basic philosophical roots of autonomy are found in Locke, Kant and Mill. Locke, in his *Second Treatise on Government*, established Man in the state of Nature to be free and equal so that none might have sovereignty over another except when freely chosen otherwise. This led to the concept of 'negative rights', the right of freedom from interference which underlies many views of democracy. Kant, in his *Groundwork for the Metaphysics of Morals*, argued that freedom is identical to autonomy and that autonomy is 'the ground of the dignity of human nature and of every rational nature'. That is, freedom/autonomy is essential to all morality. Finally, Mill, in *On Liberty*, contends that liberty is limited only by harm to others, not harm to self. This, with Locke's idea of negative rights, binds the philosophical notion of autonomy with the legal concept of a right to privacy.

Yet, taken to an extreme, these views lead to 'moral autonomism' in which each person exercises rights independently of the claims of society, and even counter to those claims. Limitations must be identified which place individuals within a moral community, with the responsibilities implied therein. That connection is made through the notion of *integrity*. Derived from a Latin word meaning wholeness or unimpaired unity, integrity encompasses autonomy while exceeding it. Usurping the capacity for self-governance violates self-governance, as does repudiating an individual's values. In these regards integrity and autonomy are synonymous. But integrity also imposes a parallel obligation to respect the integrity of other persons. That is, one cannot be whole without balancing a personal claim to autonomy with the moral claims of other persons and the community. 'Autonomy is a capacity inherent in being a rational person ... We can have degrees of autonomy ... Integrity, on the other hand, is a matter of being' (Pellegrino, 1990, p. 15).

All humans therefore have integrity – integrity of the person. This is distinct from persons of integrity. To be a person of integrity requires accommodating the moral claims presented by integrity through allowing

---

[3] This subsection draws particularly on Pellegrino (1990) with references therein.

others to act autonomously. Conversely, other parties must respect the integrity of the first party, leading to a consensual decision. Only in this way can the harmony of the individual and of society be preserved.

Viewed from this perspective, integrity, of which autonomy represents but a part, presents more of the societal context with which many cultures can be expected to share; we act concurrently with integrity so that others may be autonomous. There is nothing particularly Western about such a recognition of mutual respect and dependence.

Requirements for Autonomy in Decision Making

The medical literature identifies three factors to be considered in determining whether patients can act autonomously in making proper decisions. The three, which can apply as well to the use of genetic resources, are:

- competency;
- freedom from coercion; and
- possessing the necessary information.

Competency relating to medical treatments arises most frequently for those incapable of assessing and responding to their own self-interest. They include children, the very infirm and the profoundly retarded, among other identifiable groups. In such cases a surrogate is required, whether that be a parent, guardian, relative or, when necessary, an appointee. For the use of genetic resources such issues would generally not arise. But occasionally questions might arise if groups very isolated from late twentieth century technology could always comprehend the consequences of what is being asked of them. In such cases the communities may wish to identify an informed community representative, e.g. indigenous North Americans frequently have a tribal member, trained as a lawyer, to manage tribal property.

Coercion, while regrettably not always easily detectable, can often be countered once identified. More problematic is the question, does financial need constitute coercion? Clearly, acute need dramatically changes priorities, but, within that context, the choice made will depend on the known options. That is, even a needy individual/community can exercise some autonomy based on knowledge and competing buyers. Thus, economic coercion is at least in part a consequence of the array of buyers.

Absence of coercion would also suggest the freedom to withdraw from an agreement at any stage. This is generally allowed in the medical research context (Levine, 1975), and could be extended to research on genetic resources. The Philippines access legislation contains this provision regarding commercial research agreements (p. 58; see also Chapter 3): 'The agreement may also be revoked on the basis of public interest and welfare'. While understandable and appropriate, this clause would raise questions with a commercial entity which could see the unilateral termination of years of work. A delineation of what constitutes 'public interest', or a penalty clause which would inhibit termination in cases where the 'public interest' constituted no more than a higher paying partner once the basic work was done, would help to alleviate those concerns.

Information, its proper type and form, is the more difficult issue. Consider first the complexity of the language used. Typically, the standard in which medical PIC is 'language that the "average lay person" should be expected to understand' (Levine, 1975, pp. 3–59). Yet with genetic resources, as in medicine, requests will sometimes be directed to 'non-average' persons. Hence, the standard of language which the recipient 'might be expected to understand' would seem more appropriate. This further implies the suppliers' native language be used; clearly, the individual(s) providing the information must have a sophisticated level of use of that language as well.

The type and amount of knowledge to be provided present perhaps the most intractable problems. There are no absolute standards, but an appropriate test is suggested; can the individual demonstrate a knowledge of the relevant details of the agreement, and justify for the decision reached (Lara and de la Fuerta, 1990)? In medicine, a commonly used test of 'informed' is a two-part consent form, the first providing the information and the second containing 'a brief quiz on the essential elements of the information that have already been presented' (Levine, 1975, pp. 3–33). A similar approach might be utilized for PIC for genetic resources.

Consent in medicine is often formalized in writing, with a signature. This is useful for subsequent reference in the context of Western legal systems, but has limited relevance for illiterates (including those who use non-written languages) or in other cultural contexts. Alternatives might include video- or audio-taping the proceedings, as some demonstration that PIC has been satisfied will often be required.

Conclusions and Approaches

PIC under the CBD applies to two distinct cases, that of access to genetic resources (Article 15(5)) and biosafety (Article 19(3)). The biosafety matter compares to related issues presented in the Basel Convention (hazardous waste). A similar resolution, the specification of the competent authority to receive the information and outline of information required, would seem to serve the intent of the Convention. This would apply equally if national biosafety legislation is enacted through domestic law or via a protocol agreement. Conceptually, this matter is straightforward, even if, technically, much consideration will be required.

The broader lesson from the Basel Convention is the need for a national competent authority to rule on whether and when PIC requirements have been satisfied. This could be included in national access legislation (see also Chapters 2 and 3) as the Philippines access law establishes an Inter-Agency Committee with, among others, the responsibility to verify 'that the consent requirements in Sections 3 and 4 are complied with' for indigenous and local communities (Section 7(e))[4]. The need for a competent authority is the stronger when the PIC requirements are, as in the Philippines, cast in terms of 'obtained in accordance with the customary laws of the concerned community' (Section 2(a)). Competent authorities can further investigate if fraud or compulsion was involved in securing an agreement, in which case it should be invalidated and damages sought, as appropriate.

PIC for access to genetic resources presents broader and more complex issues, including its justification. The underlying rational is personal autonomy, a Western concept based on the eighteenth century philosophy of the Enlightenment. While Western in its expression, autonomy can be seen as a component of integrity, integrity of the person and personal integrity, which incorporates personal and societal harmony. Hence, in this broader perspective, the role of informed consent can be established. Yet it remains that those providing the information are placed in a special, elevated context of obligation to those from whom the consent is being sought.

This role is most easily appreciated when consent is viewed as a contract, and the information stage as the contract negotiation with requests for, and provision of, information. Typically, in commercial contracts, parties are expected, required for their own self-interest, to inform themselves. The same may not apply when the consent being sought is for research of no clear direct benefit to the consenter. It is in this context that medical PIC has evolved over more than a century, providing a basis of conceptualization and practical experience on which to draw for operationalizing PIC under the CBD. Consenters, for example, may wish to know the objectives of the research, which they may support or not according to their values and beliefs. The possibility of generating commercial products from the research would be one such bit of pertinent information which should be revealed through PIC.

With strictly commercial access agreements, however, the conceptual basis is quite different as the supplier is seeking specific (monetary) benefits. Expecting contractors (buyers) to provide information which could weaken their negotiating position in respect to sellers is a high standard to establish indeed. Yet law routinely does establish disclosure and information standards, which contractors are advised to read carefully before signing. For example, in the USA, when investment houses present records of past

---

[4] The Executive Order appears to contain an error in referring to Sections 3 and 4 as it is Section 2, Consent of Indigenous Cultural Communities, which establishes the basic PIC requirements.

performance, a disclaimer that past performance is not necessarily an indication of the future, is included. For PIC to be satisfied, customers need not act in accordance with information – many, for example, smoke despite the health warning mandated in numerous countries – they must simply understand the information presented, consider it adequate, and be able to justify their decision, even if seemingly irrational by probabilistic standards.

As a first guiding principle, country PIC requirements should establish different standards/requirements for commercial and research agreements. Commercial agreements may be required to include certain disclosures on intent and expected outcomes and uses. These could be reviewed and approved by a national administrative body. Unilateral termination clauses for both partners would also have to contain clauses which would penalize withdrawing solely to pursue the same work with a more profitable partner. Implementing that concept will require some careful consideration.

The provision of additional information needed by the seller to assist in striking a favourable agreement should, however, not be the responsibility of the buyer/contractor. Rather, sellers need to recognize that as being their own responsibility. In practice, some sellers will need assistance, information and capacity building to strengthen their negotiating position, at least in the initial stages. The Facilitator and other arrangements have been proposed as means of delivering that assistance on a request basis (see Chapter 2).

For primarily research agreements, those which are not intended to benefit the resource supplier directly, far more demanding PIC requirements can be justified. Suppliers are being asked to make a contribution to the public good, and should not be expected to take on the cost of informing themselves of the possible outcomes. That rather should be provided through PIC. Nor should the contractor be the sole supplier of the information and negotiator of PIC. He/she simply has too specific a commitment to securing the agreement to be expected to be completely impartial. A neutral party should also be involved, possibly as selected under national legislation, to be paid for by the would-be contractor. Suppliers should also have the right to withdraw unilaterally, although prudence suggests that right should not be made too trivially easy to invoke as much wasted time and money would be incurred.

The actual information to be provided, its language and form, can only be determined on a case by case basis. Certain guiding principles can be established, such as the provision of sufficient information for the supplier to understand the purposes of the research, and hence judge whether those objectives are to be supported or not by the personal and communal values of the suppliers. The shared information should be adequate for the supplier to predict the possible consequences. This can be verified by asking the supplier to restate the information in his/her own words and explain the decision reached. This may be done verbally (and preserved on tape as verification) or in the second of a 'two part' written consent form.

Chapter 6

Ex Situ Storage Issues Under Article 9 and the Nairobi Final Act

Of the approximately 10 million species on earth (see Chapter 1), arguably those of the most obvious importance are plants, in particular the 25 species that provide the vast bulk of human sustenance (Wilkes, 1988, p. 68). Of related importance are livestock and other organisms useful in agriculture. While all species are preferably conserved *in situ*, there are several reasons why complementing *ex situ* conservation is justified as well (CBD Article 9; see Engels, 1995, p. 17):

- Security: some *in situ* materials have been lost due to war and famine, among other causes.
- Research: easier to access for use and maintenance of relevant information.
- Cost: *in situ* storage can be complex and costly.

Hence, the CBD directs particular attention to *ex situ* storage in Article 9, as follows:

> (a) Adopt measures for the *ex situ* conservation of components of biological diversity, preferably in the countries of origin of such components;
> (b) Establish and maintain facilities for *ex situ* conservation of and research on plants, animals and microorganisms, preferably in the countries of origin of genetic resources.

For the purposes of the Convention, *ex situ* conservation is defined as (Article 2):

> conservation of components of biological origin outside their natural habitats.

And 'Habitat' as:

> place or type of site where an organism or population naturally occurs.

According to Glowka *et al.* (1994, p. 21), the term 'natural habitat' clearly and obviously excludes genebanks, and also excludes 'resources domesticated in areas other than those where they had developed their distinctive properties and maintained on farms or ranches which have not contributed to the development of those properties (for example, the fields of wheat and barley in the farms of Northern Europe)'. While the role of developed country farmers in participatory breeding, at least in past times, can be debated (Fitzgerald, 1990, described the role of 'Corn Belt' farmers in the US Midwest in the pre-hybrid era), for purposes here, 'countries of origin' shall be where those materials naturally appeared or were introduced a long time ago.

Ironically, these definitions of *ex situ*, origin, and habitat create conflicts with the role of the Convention. The intent is clearly, among other things, to promote conservation and equitable sharing of all genetic resources. Yet just those useful agricultural resources which are most important, and which society might be most willing to pay for, essentially lie outside the Convention. This is because the Convention applies only to those genetic resources which have been 'acquired [...] in accordance with this Convention' (Article 15(3); see also Chapter 2). As the great bulk of the materials held in the 1000+ genebanks worldwide (see pp. 98–99) were acquired prior to the CBD going into effect, they technically lie outside the Convention, and are not regulated by its stipulations. Exclusion from the CBD appears to follow the traditional approach of not making conventions apply retroactively (Glowka *et al.*, 1994, p. 79).

The matter of materials acquired prior to the CBD going into effect is referenced in *Resolution 3* of the Nairobi Final Act[1]:

> 4. *Further recognizes* the need to seek solutions to outstanding matters concerning plant genetic resources within the Global System for the Conservation and Sustainable Use of Plant Genetic Resources for Food and Sustainable Agriculture, in particular:
> (a) Access to *ex situ* collections not acquired in accordance with this Convention, and
> (b) The question of farmers' rights.

(The matter of farmers' rights is treated in Chapter 2, pp. 24–25).

While the referenced Global System programme is an FAO activity (see below), the COP has reiterated its intention to manage *ex situ* resources as part of the Convention mandate (see UNEP/CBD/COP/1/4, 1994 and Decision III/11). Most recently, at COP3, a willingness was expressed to

---

[1] See as well Chapter 14, 'Sustainable Agriculture and Rural Development,' of Agenda 21.

consider a Convention protocol for establishing access and use terms on the part of the Parties.

The Nairobi Final Act recommended the adoption of the agreed text of the CBD and hence is treated here as part of the obligations entered into by the signatories to the Convention. That is, there is a moral, if not legal, commitment, and practical need, to resolve issues regarding access to *ex situ* collections. That is the objective of this chapter. As groundwork, we begin with an overview of the extent and location of those holdings, and the current status of ownership and access. The focus here is on *plant* genetic resources for food and agriculture as they dominate existing *ex situ* collections and, to date, are receiving the greatest attention. Many of the issues applying to plant collections extend to other organism collections as well, but are not specifically addressed here. The overview of *ex situ* holdings is not intended to be a complete and detailed documentation, but rather to provide a perspective from which the issues of access and exchange can be better understood.

FAO (1994 data quoted in Hawksworth, 1995, Table 8.8-1) identifies over 1200 plant genetic resource collections worldwide, held in more than 160 countries and territories. These contain the estimated 4.2–6.1 million accessions (UNEP/CBD/IC/2/13, 1994, Par. 16; Hawksworth, 1995, p. 582; FAO, 1996a, p. 3-1). Many of these genebanks are small, with as few as one species and fewer than 50 accessions (e.g. Bosnia-Herzegovina, Tonga). Major collections include those of the International Agricultural Research Centers (IARCs), with 11 genebanks and apparently 600,000 accessions (Table 6.1) (FAO, 1996a, p. 3-1). Among the national collections,

Table 6.1.   IARC *ex situ* collection.

Centre	No. of accessions
ICRISAT	110,374
CIAT	70,940
CIMMYT	136,637
CIP	13,844
ICARDA	109,029
ICRAF	Data not available
IITA	39,765
ILRI	13,470
IRRI	80,646
WARDA	17,440
INIBAP/IPGRI	1,046
Total	593,191

Source: FAO, 1996a, Table 3.6.

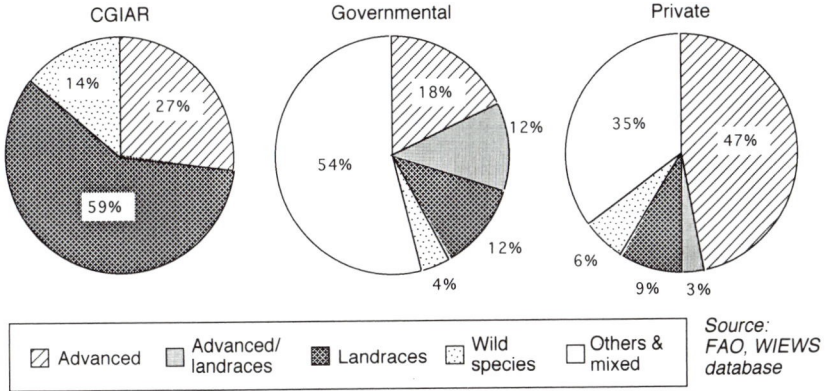

Fig. 6.1. Types of holdings by genebank ownership. (Source: FAO, 1996, Figure 3.4.)

12 countries contain over 45% of accessions held by all national collections[2], with 53% held in northern, generally non-source, countries.

Overall, governments hold 83% of accessions, the IARC 11% and the private sector only 1.27% (FAO, 1996a, Fig. 3.1). The IARC collections, however, contain about 35% of the *unique* samples, making them '... probably the world's most significant collection' (UNEP/CBD/IC/2/13, 1994, Par. 17). The nature of the holdings also differs by genebank ownership (Fig. 6.1).

Status of Ownership and Access Under the Current Multilateral System

The ownership of genetic resources can be divided roughly into pre- and post-1987 periods. During the extensive earlier period described in Chapter 2, pp. 14–21, genetic resources were treated as a 'common heritage'. Subsequently, the second period was initiated when an FAO study raised troubling questions about the ownership and control of materials held in genebanks (see UNEP/CBD/IC/2/13, 1994, Par 22; more details below). Over time, several mechanisms – three in all – have been used for access and ownership. Two of these – *unilateral* (IPR) and *bilateral* (contract) – were addressed in Chapter 2. Here emphasis is on *multilateral* systems. The assessment begins with an overview of the institutional arrangements

---

[2] The 12 countries are: Brazil, Canada, China, France, Germany, India, Japan, Korean Republic, Russia, Ukraine, the UK and the USA.

involved in multilateral systems, and progresses to a consideration of the substantive matters of ownership and access.

Institutional activities on behalf of plant genetic resources

Given the importance of plant genetic resources, a high level of institutional attention to their ownership and use is hardly surprising. Table 6.2

Table 6.2. FAO programmes on the conservation and use of plant genetic resources.

Components	Function	Status
Commission on Genetic Resources for Food and Agriculture*	Intergovernmental global forum	Established 1983 as the Commission on Plant Genetic Resources; 1967 EU members (Sept. 1996); six sessions plus one extraordinary session held; scope broadened in 1995 to include other sectors of agrobiodiversity, starting with livestock. A Panel of Experts on Forest Gene Resources is a technical advisory body to FAO
International Undertaking on Plant Genetic Resources	Non-binding agreement to assure conservation, use and availability of PGRFA	Adopted 1983; 110 countries adhere; annexes agreed in 1989 (including Farmers' Rights) and 1991. Currently under revision including for harmonization with the CBD, development of agreements on access, and the realization of Farmers' Rights
International Fund for PGR	To provide a channel for support and promotion of sustainable PGR conservation and use at a world level	Not yet operational. Principle agreed by FAO Conference; Global Plan of Action (GPA) will be useful in determining requirements for Fund
Global Plan of Action for the Conservation and Sustainable Utilization of PGRFA	To rationalize and improve the international efforts for the conservation and use of PGRFA	First plan adopted by International Technical Conference (TIC) on PGR, in June 1996
Report on the State of World's PGRFA	To report on all aspects of conservation and use of PGRFA to identify gaps, constraints and emergencies	First report adopted by International Technical Conference on PGR in June 1996

Table 6.2. Continued.

Components	Function	Status
World Information and Early Warning System (WIEWS)	To collect and disseminate data on PGRFA and related technologies; identifying hazards to genetic diversity	Information system established, including records of *ex situ* collections in 135 countries. Early Warning System at planning stage
Network of *ex situ* Collections under the Auspices of FAO	To facilitate access to *ex situ* collections on fair and equitable terms	Established with collections of 12 IARCs (agreement signed in October 1994); 31 countries expressed their willingness to include their collections; one has signed agreement. International standards for genebanks agreed

* Total number of countries and regional economic integration organizations which have become members of the CGRFA and/or adhered to the Undertaking is 149.
Source: FAO, 1996a, Table 6.4.

contains a synopsis of the major activities underway in this regard. Much is organized by the FAO, the United Nations' specialized agency for food and agriculture. The FAO has had activities regarding plant genetic resources since its creation in 1945. The passage of the CBD with its inclusive scope for all genetic resources raised some coordination and priority issues regarding the FAO programmes. Both the COP and FAO have endorsed the adoption of the International Undertaking (Decision II/15) and encouraged its rapid revision to place it in harmony with the CBD (Decision III/11). Some observers believe that harmonization will require 'some real, possibly "radical" reform' and that the FAO Commission on Genetic Resources for Food and Agriculture must be 'greatly intensified' to achieve the revision by COP4 in 1998 (*Diversity*, 1996a, p. 8). Steps to date are identified below.

The formal FAO activities are structured around the International Undertaking on Plant Genetic Resources (IUPGR; the Undertaking), a non-legally binding agreement adopted in the FAO Conference in 1983 (Resolution 8/83). It constitutes the only instrument for FAO regarding plant genetic resources. That Resolution contains six articles pertinent for the issues under consideration here, as follows:

- exploration and collection of genetic resources (Article 3);
- conservation *in situ* and *ex situ* (Article 4);
- availability of genetic resources (Article 5);
- international cooperation in conservation, exchange and conservation (Article 6);

- international coordination of genebank collections and information systems (Article 7); and
- funding (Article 8).

Three resolutions were subsequently adopted and incorporated as annexes. These include: (i) an agreement that PBR were not inconsistent with the Undertaking; (ii) recognition of Farmers' Rights; and (iii) a reaffirmation of the sovereign rights of nations over their genetic resources (see discussion in Chapter 2). Subsequently, Resolution 7/93 called for the revision of the Undertaking in harmony with the CBD as noted above.

Also, in 1983, on the recommendation of members, the FAO began developing a comprehensive Global System for the Conservation and Utilization of Plant Genetic Resources for Food and Agriculture (the Global System) which, among other things, established the Commission on Plant Genetic Resources for Food and Agriculture. Table 6.2 highlights the status of the Global System. It is one of the roles of the Commission on Genetic Resources to 'facilitate and oversee co-operation between FAO and other governmental and non-governmental bodies [...], in particular with the Conference to the Parties to the CBD ...' (Terms of Reference, FAO Report of the Council, 110th session).

A recent step towards harmonization was the FAO 4th International Technical Conference (the Leipzig Conference) held in June 1996. The Leipzig Conference had the dual objectives of: (i) deliberating on the State of the World's Plant Genetic Resources (report in draft form, FAO, 1996a) leading to the adoption of the Global Plan of Action (GPA), and (ii) adopting the *Leipzig Declaration on Conservation and Sustainable Utilization of Plant Genetic Resources for Food and Agriculture*. The draft GPA contained 346 recommendations distilled from the draft report formulating guidelines for future funding, conservation, and use. About 95%[3] of the items were adopted under four 'Priority Activities': (i) *in situ* conservation and development; (ii) *ex situ* conservation; (iii) utilization of plant genetic resources; and (iv) institutions and capacity building. But not so the matter of the use of materials subject to property rights nor the final section, 'Ensuring a Fair and Equitable Sharing of Benefits', which, by one account, incorporated an effective implementation of farmers' rights (Pistorius, 1996). However, the US government, a long-time critic, did support the 'global plan calling upon the world community to support farmers' rights' (Putterman, 1996). Yet the terms chosen – '*acknowledge* the roles played' and '*desirability* of sharing equitably benefits' (emphasis added; see Box 6.1) – are non-committal and follow closely the wording of the CBD, so there is a question if any real additional endorsement has been added. As regards the funding arrangements, notably for *ex* and *in situ* conservation, an agreement last

---

[3] Of the 346 draft recommendations, 325 were approved for inclusion in the GPA.

year 'de-link[ed] the adoption of the Global Plan from its funding' (Mooney, 1996). Despite the clear focus of Southern representatives in the availability of new and additional monies, Northern representatives had no mandate to negotiate financial terms, leading to a postponement to a later meeting where it will be discussed as regards industrialized countries and multilateral agencies, in close cooperation with the CBD (Nemoga-Soto, 1996). Overall, the International Institute for Sustainable Development (1996) noted, 'a strong lack of commitment to the Plan, particularly by some key donor countries [...] over a commitment to cover the full incremental cost of implementing the GPA'. Relevant parts of the Leipzig Declaration are included in Box 6.1.

Box 6.1.   Leipzig Declaration on Conservation and Sustainable Utilization of Plant Genetic Resources for Food and Agriculture – selected aspects associated with access and benefit sharing.

1.   ... recognizing the desirability of sharing equitably benefits arising from the use of traditional knowledge, innovations and practices relevant to the conservation of plant genetic resources ...

2.   Recognizing that states have sovereign rights over their plant genetic resources ...

4.   We acknowledge the roles played by generations of men and women farmers and plant breeders, and by indigenous and local communities, in conserving and improving plant genetic resources.

5.   ... many [genebanks] cannot meet minimum international standards. An alarmingly high number of stored accessions are in need of regeneration ...

6.   The critical linkage between conservation and utilization should be improved.

7.   Access to and the sharing of both genetic resources and technologies are essential ... Such access to and sharing of technologies with developing countries should be provided and/or facilitated under fair and most favorable terms, including on concessional and preferential terms, as mutually agreed to by all parties to the transaction ... terms which recognize and are consistent with the adequate and effective protection of intellectual property rights.

8.   It is important that the diversity be made more useful and valuable ... by providing better and more accessible documentation.

9.   Means are needed to identify, increase, and share fairly and equitably the benefits ...

10.   ... we have adopted the Global Plan of Action for the Conservation and Sustainable Utilization of Plant Genetic Resources for Food and Agriculture.

Source: FAO, 1996b, App.1.

Ownership and access

The 1987 legal opinion of FAO noted that 'ownership of genetic material held in government genebanks or those of public institutions was, in most cases, for practical purposes considered to be vested in the State in which these genebanks are located. However, for material held in the International Agricultural Research Centers (IARCs) the legal position was unclear' (UNEP/CBD/IC/2/13, 1994, Par. 22). This state of affairs was considered unsatisfactory by the Commission on Plant Genetic Resources which in 1989 called for the implementation of Article 7.1(a) of the International Undertaking (Box 6.2).

Box 6.2. FAO International Undertaking Articles Pertaining to an International Network of Genebanks.

Article 7.1(a): There develops an internationally coordinated network of national, regional and international centres, including an international network of base collections in genebanks, under the auspices or jurisdiction of FAO, that have assumed the responsibility to hold, for the benefit of the international community and on the principle of unrestricted exchange, base or active collections of the plant genetic resources of particular plant species.

Article 7.2: Governments or institutions [...] may, furthermore, notify the Deputy-General of FAO that they wish the base collection or collections for which they are responsible to be recognized as part of the international network of base collections in genebanks, under the auspices or the jurisdiction of FAO. The centre concerned, whenever requested by FAO, will make material in the base collection available to participants in the Undertaking, for purposes of scientific research, plant breeding or genetic resource conservation, free of charge, on the basis of mutual exchange, or on mutually agreed terms.

By mid-1996, 32 countries had expressed a willingness to join the International Network of Ex Situ Collections. In 1994, the IARCs signed an agreement placing most of their collections in the International Network. Finally, member institutions of the IPGRI Register of Base Collections, about 50 institutions which have agreed to long-term conservation and unrestricted availability of specified germplasm, will consider joining the International Network. If completed, the network would cover about 70% of global accessions (UNEP/CBD/IC/2/13, 1994, Par. 25).

The International Network functions on the basis of *designated germplasm* which is voluntarily placed in the Network. Hence it is possible for a single facility to have materials available on distinctly different conditions based, in part, on the date the materials were received. The CIMMYT and

Box 6.3.   CIMMYT and IRRI policies on intellectual property.

CIMMYT
CIMMYT genetic resources held in trust

1.   The genetic resources accessed prior to the ratification of the Biodiversity Convention (29 December 1993) and identified accession numbers in the CIMMYT genebanks are held in trust for the world community. CIMMYT will not seek intellectual property protection of these plant genetic resources and is opposed to the application of protection legislation to them. For these materials, CIMMYT adheres to the principle of unrestricted availability ...

2.   The availability of genetic resources received after the ratification of the Biodiversity Convention will be governed by the conditions defined by the provider ...

Products of research

1.   CIMMYT will protect or allow others to protect CIMMYT's intellectual property only when we see this as compatible with our mission.

[Author's note: Available information on the germplasm is provided on equivalent terms.]
Source: CIMMYT, 15 April 1994.

IRRI
Principles

1.   IRRI will continue to make genetic resources that it holds in trust freely available ...

2.   ... IRRI will supply these genetic resources under a Material Transfer Agreement (MTA) designed to ensure the free availability of the materials and of genes derived directly from them.

3.   The genebank may accept materials for 'black box storage' on which the donor of the germplasm has placed distribution restrictions only when refusal to accept such restrictions would endanger the long-term preservation of such germplasm.

4.   All breeding materials, elite germplasm, and parental lines of hybrid rice that are derived from conventional breeding will be made freely available. However, an MTA will be used for the exchange of these materials with private sector organizations.

5.   Where it is essential to make advanced biological technologies available to developing nations, IRRI may accept limitations on distribution of the derived and associated materials.

6.   To ensure availability to developing nations of advanced technologies or biological materials such as microbial strains, IRRI will, exceptionally, apply for intellectual property protection ...

Source: IRRI Policy on Intellectual Property Rights.

IRRI policies on intellectual property specifies access terms (Box 6.3). These terms are reflected, for example, in CIMMYT's MTAs.

The International Network with its designated germplasm, combined with the IPR policy, establishes the parameters of the current access system, which has been dubbed as the informal multilateral system. This is the system which is often perceived as being 'no longer broadly acceptable' (IPGRI, 1996a, Par. 27; see also Cooper *et al.*, 1994). From the supplier's perspective, this system does not provide the opportunity for monetary benefit, while some less experienced users do not understand the requesting procedure, rendering it 'non-transparent.' This system is none the less responsible for providing without charge approximately 800,000 accessions annually, 650,000 from the IARC genebanks alone, the bulk requested by developing countries (IPGRI, 1996b).

Future Plans

Because of the concerns over the acceptability of the current internal multilateral system, FAO in 1995 commissioned IPGRI to conduct 'an in-depth factual study on the various possible technical options [...] for a multilateral system that would facilitate access to genetic resources for food and agriculture on mutually agreed terms while offering ways and means of equitable and fair sharing of benefits arising from the use of these genetic resources in commercial and/or non-commercial purposes to developing countries (countries of origin) ...' (IPGRI, 1996a, Annex II; herein, the Report). Due attention was to be placed on keeping transaction costs low so as not to burden governments, donors, or, possibly, users.

The Report identified three basic approaches, as follows:

- current informal multilateral approach ('open exchange');
- strictly bilateral approach (contracts); and
- a multilateral framework governed by mutually agreed rules (dubbed 'MUSE' for MUltilateral System for Exchange).

For purposes here, the bilateral approach was assessed in Chapter 2, and the current informal system in the previous subsection. Hence, attention here is on the MUSE. The bulk of the Report itself is focused on the MUSE which appears to have led some readers to believe that is the preferred outcome despite the professed neutrality of IPGRI.

The MUSE itself carries over aspects of the current informal system, but with '... a more formalized approach to exchange, although one that retains the essential features of the current system' (IPGRI, 1996a, Par. 44).

This combination of flexibility and formality is to be accomplished through a series of options, including possibly:

- optional bilateral benefit sharing for selected materials/special conditions;
- mechanism for financial compensation;
- access assistance, possibly through a secretariat, plus a detailed database/information service; and
- agreement among members establishing standard rules for access and prior informed consent.

Still other issues are left open, including terms of membership (governments are proposed to establish rules/benefits for national non-governmental members), funding for expenses (possibly a membership fee), and scope. Scope refers to the range of biological material available through the system. Several definitions of scope are proposed, indicating at a minimum just how complex a matter this can be (IPGRI, 1996a, Chapter 4):

- scope defined comprehensively;
- scope defined by taxon or genepool;
- scope defined by category (e.g. landraces, wild relatives);
- scope defined by date of collection (pre/post CBD);
- scope defined by conservation method (*in/ex situ*);
- scope defined by ownership; and
- scope defined by intended use, and scope defined by mixed options.

Compelling arguments are made in the Report as well as elsewhere (Cooper *et al.*, 1994) that plant genetic resources are public goods. From a more practical perspective, the vast bulk of material appears to have limited commercial value. A high transaction cost system could easily absorb any value, meaning complex processes/hindered access with no residual benefit sharing, a most undesirable outcome. That aspect was brought forward more clearly in the Options report containing cost estimates of the multiple possible components of a system or systems (IPGRI, 1996b). Key cost components, reported in Table 6.3, should be considered preliminary at this stage but are indicative of the relative costs of different approaches to sharing.

These cost estimates, although preliminary, do suggest strongly that some options are impractical, at least with current technology. This includes 'genetic fingerprinting' for all materials, and multiple court challenges for enforcing agreements (although arbitration may be feasible). Even the tracking of agreements is costly, meaning it is not practical for most material. Note should also be made that many of the costs are fixed: a bilateral system which reduces the number of accessions distributed increases the costs allocated on a per-accession basis.

Support in the form of a database and central contact to identify the availability of material is moderately expensive, and feasible, but the need

Table 6.3. Cost component estimates for alternative distribution systems for plant genetic resources.

	Costs in US$
Collecting (accessible seeds only; other materials higher costs)	$10–30/accession
Conservation (variable costs only)	$50/accession
Regeneration (seeds only; vegetatively propagated more costly)	$50–350/accession
Multiplication (varies by crop)	$2.00/accession
Distribution of sample	
Postage	$0.25/accession
Phytosanitary permits	$0.50/accession
Global information network (hardware and clean up data files) 50 institutions @ av. $50,000/institute	$2.5 M
Negotiating exchange agreements	$100 (simple)
	$1000 (complex)
Trading and monitoring use	
Limited	$1800/agreement
Systematic	(homogenous)
	$170–500/accession
	(heterogenous)
	$17,000–10,000/accession
Enforcing agreements	
Litigation	$15,000–300,000/case
Arbitration	$5000–21,500/case
Establishing secretarial office	$250,000/year
Establishing and managing compensation fund	
Office	$250,000/year
With board added	$100,000/year
Negotiating agreement	$330,000–500,000

Source: Lesser, 1997.

should be firmly established before making the commitment. Internet technology may soon render outmoded the need for a centralized access system, which all too likely could become an incumbrance to the flow of germplasm.

Hence, a general bilateral system involving efforts to enforce all agreements would be economically infeasible. The MUSE, by rendering those costs unnecessary, then becomes attractive. The MUSE, however, retains the key weakness of the current informal system: there is no direct payment mechanism for financially compensating suppliers of materials designated for open exchange. Brush (1996a) notes that the *status quo* appropriation system does provide for some indirect reciprocity in the form of information and improved varieties, but that is clearly considered as inadequate by genetic resource suppliers. The 'compensation fund' proposed in the MUSE Report skirts the Farmers' Rights fund issue too closely for any real confidence in its enactment (see Chapter 2). This presents a dilemma indeed; common property provides general benefit which should be compensated for at general cost. But those selected for providing the compensation (developed countries and/or private companies) have been unwilling to do so.

Part of this unwillingness has been institutional: a 'lack of confidence in management at the FAO' (Shands, 1994). Yet more than revised management would be required for enacting a fund. Part is the absence of any real discussion of how the funds would be used; I have argued elsewhere that the utilization of the funds is more difficult than its management, yet fundamental to a willingness to contribute. A detailed dispersement system in itself, though, would not bring an operating fund forward. A fundamental issue, before and with MUSE, is how to provide incentives for suppliers to participate in exchange, or to 'designate' germplasm, to use the current terminology.

The FAO Commission on Genetic Resources for Food and Agriculture at its December 1996 meeting reviewed the MUSE Report and Options document and decided to proceed with the process of considering the options in a systematic manner (the 'matrix' approach) and in consultation with Commission members (*Diversity*, 1996a). A final resolution to the complex scientific and political matter is likely to be some time away and (in the author's judgement) is likely to maintain a substantial multilateral component, if only on deference to the costs of managing bilateral agreements. At the same time, more of the costs of storage and replication (with overhead) are likely to be assessed on an ability-to-pay basis.

Conclusions and Approaches

Beginning in the 1920s, and accelerating following the establishment of the IARCs beginning in 1960, a reasonable *ex situ* collection of plant genetic resources for food and agriculture has been assembled. The tradition throughout this period has been for coverage of costs by governments and, later, bi- and multilateral donors. Treated as a common heritage with no charges or limitations imposed, access and use have been excellent.

Strains on that system began to be evident in the 1980s through the supplying countries, no longer satisfied with a common heritage system which provided no direct compensation. Indirect compensation, provided by the system in the form of knowledge and some improved varieties, clearly was not considered adequate either.

As part of this process, a significant (and, to the outside observer) complex institutional mechanism has been established by FAO. That system produced the concept of Farmers' Rights and a Global Plan of Action. While there have been broad agreements on the details and the general objectives, the partitioning of materials into open (designated) and restricted access materials, these efforts, one after another, have been stymied over the question of who pays whom, and for what. The current version, the 'MUSE' Report, suggests a number of alternative flexible, yet standardized, approaches, but does not resolve the payment conundrum. The cost figures in the Options Report are, however, convincing on why a general bilateral system will not work well; costs are simply too high. Thus, repeatedly, the same question emerges: what is the specific (e.g. monetary) incentive for suppliers to participate in an open access, multilateral system?

Potential funding sources are easy to identify, either a general fund (from bi- and multilateral, with possible private sector) donors, and users. The general fund has not worked, and there seems no reason to believe at this point it will work; it may, but there is nothing to count on. At the same time, there are strong conceptual reasons, backed by decades of traditional practice, arguing against placing a charge on the materials themselves. Indeed, it is not clear that the IARCs could charge, given the vague 'in trust' status under which much material was transferred.

In my view, one way out of this seeming dead-end is to *separate access to the materials from access to the information regarding it.* The material itself belongs as a common heritage resource. Included would be 'passport data' identifying where it was collected. The material, however, has limited value with limited information; identifying its useful characteristics requires a considerable investment in growouts and/or testing. Presently, a major investment in screening is typically made only when no known material has, say, a desired disease resistance. That and all other available information is now distributed without charge with all accessions. One consequence is that most entities have no incentive to invest in screening, and much useful material is never identified and used. 'Typically most genebank accessions have not been well characterized and evaluated, a situation that leads to the under use of collections and failure to realize their full value ...' (FAO, 1996b, Par. 145).

A system that allowed charges for information would provide incentives for national programmes to add value through screening trials. This could be done without involving the complexities of defining ownership terms or the need to limit replicates in multiple collections (and thereby endangering conservation). If the information sold came all or in part from a local or indigenous community, a portion of the value could be returned to them. (The indigenous property rights system proposed in Chapter 7, pp. 128–133, includes a similar separation of ownership of the materials from use information.) Finally, a value of the information can be impartially calculated as the cost of its generation. The price need not be set equal to that value, but it provides a baseline and would be expected to be lower than the cost as: (i) costs can be recovered from multiple buyers and (ii) charging too high a price will lead to recreating rather than purchasing the information.

Sellers can charge multiple prices, say one to developing countries, one to developed ones, and a third to the private sector. Or information can be used in barter, for other information or training. Differential pricing could help overcome the cost burden which would be placed on developing country breeders who currently pay nearly nothing.

The issue none the less remains – what incentive exists for national programmes to donate their genetic resources for public use while selling associated information? Programmes after all could sell both the information *and* materials. Limiting sales to information will require an incentive. Several possibilities exist.

Donors could underwrite screening costs on the condition genetic resources are designated. This would benefit suppliers immediately through the provision of training and longer term through possible sales, while expanding the usefulness of materials for breeders. In short, rather than transferring monies from one group to another (the underlying function of the compensation fund), this approach *creates* value. The CGIAR SINGER (System-wide Information Network on Genetic Resources) Project could assist in selling the information on designated materials by including a marketing system in its developing global information system (see FAO, 1996a, Chapter 6). In all, marketing costs will be less than that of the germplasm itself.

Several limitations are, however, apparent. One is the absence of a direct connection between the payments and conservation. But then there is no clear connection between payments for the *ex situ* germplasm and *in situ* conservation either. Second, no immediate payment flow is possible as the marketable information remains to be generated. This limitation could be overcome in part by donor support for training and structuring the screening process. Perhaps most significant is the protection of the information provided. If sold to a single user, what assures that it will not be passed on to others, limiting the revenue possibilities? This situation is directly parallel to the issue of controlling the unauthorized use of genetic resources, but in ways information is easier to share and more difficult to trace.

Several possible, if incomplete, remedial actions exist. Owners of information can *copyright* it, providing a legal remedy should unauthorized use occur. Copyright, unlike patents, is easy and inexpensive to secure in multiple countries, with most countries having a form of copyright protection available. Detection, not remedies, though remains the underlying problem. Suppliers could follow the strategy of giving the germplasm associated with their information a unique code name for each purchaser. A subsequent request by other than the information purchaser to a genebank for that named material would indicate an unauthorized use, with penalties specified for both the user and supplier of that information. Information suppliers could also establish a technical support service available only to purchasers of information, increasing the value of purchasing (as opposed to acquiring) the information. The scientific issues involved with the production of that service, and hence its practicality, exceed the scope of this volume.

In the final analysis, genetic resources fit well the concept of common property. Converting common property to private property for which payment may be collected is complex and costly in transaction costs. Privatizing the information, as opposed to the materials themselves, while not avoiding all the problems, does simplify the process while providing incentives for increasing information as opposed to restricting information flows. It should be given serious attention by suppliers and donors alike, particularly if the idea of an access protocol is advanced.

Chapter 7

Roles and Treatment of Traditional Knowledge Under Articles 8 and 18

There is an oft-repeated statement in the CBD-related literature to the effect that cultural and biological diversity are intertwined, '... the preservation and maintenance of biological diversity goes hand-in-hand with the preservation and maintenance of cultural diversity' (UNEP/CBD/SBSTTA/2/7, 1996, p. 11; see also McNeely, 1995). Hence, the emphasis on the preservation and utilization of traditional knowledge, a product of cultural diversity.

Despite this level of attention and importance accorded local/indigenous (L/I) knowledge, there is no succinct, generally accepted definition of what it is. Brush (1996b, p. 4) identified two definitions of indigenous knowledge. The broader one refers to 'systematic information that remains in the informal section'. 'Indigenous knowledge is culture-specific, whereas formal knowledge is decultured'. The narrower definition – which Brush contends is the more common usage – refers to the 'knowledge systems of indigenous people and minority cultures'. In these ways, L/I knowledge is typically defined in the negative, as a contrast with 'Western scientific knowledge', which knowledge is perhaps best characterized as being systematic, subject to the scientific paradigm of hypothesis testing ('proof').

Conversely, under the 'scientific method', even intuitively apparent relationships cannot be accepted until proven. For perhaps a century, this scientific approach was widely accepted as the leading if not only path to knowledge, in part because of successes in disease treatment and prevention. Through such an extreme elevation of this approach alone, other forms of knowledge were necessarily marginalized, and much valuable wisdom has been lost. Moreover, the approach has been criticized as 'reductionist', reducing nature to mechanical processes and thereby ignoring the key social and cultural aspects (Shiva and Moser, 1995).

More recently, there seems to be a worldwide shift, or at least enlargement, of attitudes about sources of useful information (Muchena and Vanek, 1995, provide a brief sketch of this transformation in attitudes as applied to ecology). In particular, scientists are treated with greater scepticism, in part due to the all-too-apparent problems associated with much modern technology, from pesticide-contaminated drinking water to the labour-displacing effects of automation. In its place is a broader appreciation of more conventional wisdom, including, to a larger degree in the West, non-traditional medical therapies. Indigenous peoples in particular, who 'live in intimate contact with their major resources', provide the 'kind of contact with nature' that the majority of us specialized, urbanized inhabitants have lost (McNeely, 1995).

Sometimes, the positions taken are extreme, romanticized, and unsubstantiated, such as how the current levels of world food production could be maintained using technologies from the last century. Moreover, there is much in the natural world which is not detectable by the (largely) unaided human senses, so much lies beyond the scope of L/I knowledge systems (Bentley, 1993). Conversely, much useful in life is not measurable, and so lies outside the scientific method.

L/I knowledge can be seen in contrast to the 'scientific method' by noting it is evolutionary, community based, undertaken for tangible benefits, and based on observable benefits within the community where developed. Such knowledge is empirical in a pragmatic sense with cultural and/or spiritual relevance unique to communities. Thus, it may not have utility beyond that community, but on occasion does. Shaman Pharmaceuticals, a 1989 startup, relies on indigenous usage for the initial identification of plant prospects for developing medicines, but considers a lead as occurring only when three separate communities use the same plants (Asebey and Kempenaar, 1995; see also Chapter 2). Yet, even localized, L/I knowledge is invaluable in ecosystem management, which brings us back to the CBD.

In the terms of the CBD, there are three aspects of achieving the hegemony of the preservation of cultural and biological diversity:

- respect, preserve and maintain knowledge, innovations, and practices;
- promote use, including through cooperation; and
- encourage equitable sharing of benefits.

The relevant articles are 8(j) and 18(4), as follows (emphasis added):

> **Article 8(j):** Subject to its national legislation, *respect, preserve and maintain knowledge, innovations and practices* of indigenous and local communities embodying traditional lifestyles relevant for the conservation and sustainable use of biological diversity and *promote their wider application* with the approval and involvement of the holders of such knowledge, innovations and practices

and *encourage the equitable sharing of the benefits* arising from the utilization of such knowledge, innovations and practices;

Article 18(4): The Contracting Parties shall, in accordance with national legislation and policies, encourage and develop methods of *cooperation for the development and use of technologies, including indigenous and traditional technologies,* in pursuance of the objectives of this Convention. [...]

The CBD's first requirement for L/I knowledge, the according of *respect* is, in ways, the broadest and most difficult, encompassing as it does: (i) 'secure tenure to [ ] traditional lands', and (ii) a 'status comparable to that shown to other types of knowledge' (UNEP/CBD/SBSTTA/2/7, 1996). The former is undoubtedly the most fundamental, but exceeds the scope of this volume. Indeed, and, regrettably, it may exceed the scope of the CBD which, through its United Nations roots, functions in conjunction with the sovereignty of national governments (*Diversity,* 1996b). It appears some of the attention given to indigenous knowledge under the Convention is a mechanism for publicizing the plight of many of the world's indigenous peoples. The latter point, the status of the knowledge, is predominately a matter of public opinion which, while changing, evolves slowly. There is reason to doubt if legislation would serve a function in this regard, as is proposed by some (see UNEP/CBD/COP/3/22, 1996, Par. 37). Overall, this issue too exceeds the scope of this volume.

Where attention *is* focused here is on aspects of *promoting wider application* and *equitable sharing* of benefits. Specifically, the subject of this chapter is mechanisms for claiming ownership to L/I knowledge. Without such mechanisms, knowledge must be protected by absolute secrecy or it becomes part of the public domain. Explicit ownership systems for their part allow both control, possibly including refusal to share when use would be counter to community norms, and a basis for negotiating equitable sharing. A challenge is how to provide for the commercialization of L/I knowledge 'without destroying the continuation of the social structures that have generated this knowledge and on which the livelihoods of many indigenous communities depend' (UNDP, 1995, section B, Par. I.8). Calls for a property rights (a broader concept than the formal means of IPR) system most certainly did not originate here, nor from NGOs speaking on behalf of L/I communities, but from within indigenous organizations themselves. Samples are included in Box 7.1 (quoted in UNEP/CBD/SBSTTA/2/7, 1996, and UNEP/CBD/IC/2/14, 1994, with sources).

The kinds of issues underlying the deep concerns expressed in the COICA Statement are reflected for many in the W.R. Grace Co. patenting (in the USA) a product of the neem tree without returning any of the benefit to local Indian communities which long used the nuts in a way related to a pesticide. Jeremy Rifkin of The Foundation on Economic Trends, representing a large group opposed to that patent in particular, has filed a

Box 7.1.   Indigenous community organization positions on IPR.

Charter of the Indigenous-Tribal Peoples of the Tropical Forests (1992)

Article 44: Intellectual Property: ... we demand guaranteed rights to our intellectual property, and control over the development and manipulation of this knowledge.

Kari-Oca Declaration of Indigenous Peoples on Environment and Development (1992)

Paragraph 102: ... we require that our rights to intellectual and cultural properties be guaranteed and that the mechanisms for each implementation be in favour of our peoples and studied in depth and implemented.

Mataatua Declaration on Cultural and Intellectual Property Rights of Indigenous Peoples (1993)

1.1   Define for themselves their own intellectual and cultural property.
1.2   Note that existing protection mechanisms are insufficient for the protection of Indigenous Peoples Intellectual and Cultural Property Rights.
2.6   Indigenous flora and fauna is inexplicably bound to the territories of indigenous communities and any property rights claims must recognize their traditional guardianship.
2.8   A moratorium on any further commercialization of indigenous medicinal plants and human genetic materials must be declared until indigenous communities have developed appropriate protection mechanisms.

Coordinating Body of Indigenous Organizations of the Amazon Basin Statement (COICA) (1994)

I.1   Prevailing IPR systems reflect a conception and practice that is:
colonialist, ...;
racist, ...;
usurpatory, ...
I.10   Patents and other IPR forms applied to life forms are unacceptable to indigenous peoples.
I.13   Prevailing IPR systems must be prevented from robbing us, through monopoly rights, of resources and knowledge ...
II.5   Study the feasibility of alternative systems and mechanisms for protection of indigenous interests ...
II.6   Seek to make alternative systems operational within the short term ...

petition in the US Patent Office to have the patent withdrawn (*Biotech Reporter*, 1995). For its part, Grace argues that the patent is for a new formulation of neem which has a longer shelf life, extending access to this biopesticide worldwide. This, in the short term, has increased the value of neem nuts, benefiting collectors at the expense of users (see Kocken and van Roozendoal, 1997). For the longer term, plantations are being established, starting a new industry in several countries. The neem itself, native to India and Burma, has been successfully spread to West Africa, Saudi Arabia and the Caribbean, among other countries/regions (National Research Council, 1992).

Means of Protecting Traditional Knowledge

Brush (1993) identifies three approaches for the compensation of L/I knowledge, as follows:

- top-down approach with international and national agencies extending rights to L/I knowledge;
- middle-ground approach utilizing existing IPR; and
- bottom-up approach with L/I communities claiming rights.

Here we are concerned with the latter two options.

Traditional intellectual property rights

There is no reason why L/I innovations could not occur in any technological area; human ingenuity has no known limits. However, as a practical matter under the Convention, references here to IPR are to genetic resources and to conservation-enhancing technologies, such as bioremediation. Hence, it is desirable to focus here on the applicability of IPR to genetic resources, including landraces, as major areas for L/I technology development. Such subject areas, of course, include the knowledge associated with those materials. IPR are also discussed in Chapter 2 as regards genetic materials ownership and in Chapter 9 relating to technology transfer. Here the focus is on ownership of knowledge, whether embodied in genetic materials or not.

Patents

Patents may be sought for genetic resources in the form of the entire organism (microorganism, plant) or parts thereof, such as a gene complex, provided there is some human input. In general, the patenting of genes (except human genes) is not a legal problem; indeed, it is not entirely clear they would be treated as living organisms. Similarly, many countries allow patents for microorganisms, and Trade-Related Aspects of IPR (TRIPs) under GATT/WTO (see Chapter 9, pp. 175–176) specify that microorganisms may not be excluded from patent protection. Seeds/plants are a more complex matter, and animals yet more so.

Seeds are patentable subject matter in the USA and provisionally elsewhere. There is no inherent reason why genetic materials of agricultural, pharmaceutical, and other uses would not likewise be patentable, at least in concept. The fact that the materials are identified in the wild rather than purposely invented is itself not a legal hindrance. Precedence has been established with patenting microorganisms identified in the wild as long as the application is in a 'culturally pure' form to reflect human intervention (see Bent *et al.*, 1987; for the situation in Europe, see Straus, 1994). Indeed, what is really being protected is the human knowledge of how the organism is to be used. The other patent requirements must be fulfilled as well. Nor does the fact that (when required) regenerating a plant *de novo* from the technical description is difficult and expensive, if possible at all, pose an absolute barrier to patenting under the disclosure requirement. In such cases, a deposit is generally required (Straus and Moufang, 1990). Thus there is nothing fundamental which prevents the patenting of these materials where seeds and plants in general are patentable. Much the same conclusions can be reached for animals, although the technical issues are often more complex.

The hindrance is rather a practical matter. Patents are not granted for a plant in its entirety, but for a plant (or other product) with unique characteristics, as specified in the patent claims (US Department of Commerce, 1983). For plants in the past those attributes have been elevated tryptophane levels, herbicide resistance, and the like among agricultural applications; attributes introduced/induced through technological procedures. It is likely some landraces have such unique attributes – one traditional potato variety for example had hairy leaves which aided in aphid (and hence virus) resistance – but certainly not all. For pharmaceutical and industrial applications, generally a genetic sequence is identified and removed from the source organism. Identifying and characterizing such traits at the level required by patent offices is a significant task, certainly beyond the means of local communities and, given the particular requirements of patent applications, exceeding the expertise in many countries. A final consideration is the cost of preparing an application, about $US20,000 for a US application and twice that in Europe (due to translation charges) (Abbott, 1993)[1]. While it may be possible to locate funding for processing some patent applications, that procedure would not be feasible for large numbers of materials which had not been carefully screened, implying a very low probability of commercializable products (see Weiss, 1995; Principe, 1991). Thus patents

---

[1] Costs in Europe were recently reduced by 20% and delayed the payment of national designation fees, but major translation costs remain (Meller, 1997).

are not practical for protecting genetic materials in bulk, although they may be used in certain cases, where permitted.

Another category of patents with some useful attributes is petty patents (alternatively called utility models). Petty patents are in effect a weaker form of patent for more modest inventions. They are distinct because the duration is typically up to 10 years as opposed to around 20 years, and the standard for the invention (the inventive step requirement) is typically lower. Thus applying for and receiving a petty patent is generally less expensive than for a full patent, although the royalty rate would as a result be expected to be lower as well. The Japanese system has the added option of switching from a petty to a regular patent application. That provides additional flexibility. Studies of petty patent systems indicate that they are effective in encouraging investment at the local level in developing countries (Evenson, 1994; Evenson *et al.*, 1987).

The principal limitation with petty patents is that they are usually designed for and specifically limited to manufacturing products. The Japanese utility model law for example reads 'shape or construction of articles or combination of articles so as to contribute to the development of industry' (Law No. 123, 1959, Section I.1). For developing countries, a plough design would be an example. Kenya is an example of an innovative system where petty patents have recently been allowed for traditional medicinal knowledge (Gollin, 1993). That system should be studied for possible application elsewhere.

Plant Breeders' Rights

Plant Breeders' Rights as embodied in the International Union for the Protection of New Varieties of Plants (UPOV) are a form of patent-like protection expressly for plants. There are some technical differences which as a general matter make the extent of protection less broad than for a patent. However, PBR, while referring to one or more distinct attributes of the plant, apply to the whole plant. PBR are relatively easy and inexpensive to apply for, costing about one-tenth the amount of a patent (Plowman, 1993). PBR include a specific research exemption which provides good access to protected materials. Furthermore, varieties discovered in the wild are protectable with PBR, although some breeding would typically be required to satisfy the homogeneity and stability requirements (Straus, 1988; Juma and Ojwang, 1989). Hence, PBR would seem to apply to many of the needs for protecting genetic materials in agriculture. UPOV is not intended to protect plants in general as is made evident from the list of genera to be protected under the 1961 Act (Article 4(3)). For example, it would not generally be applicable to wild plants used for pharmaceutical purposes.

Where PBR fail even for agricultural uses, or would seem to fail, is in not providing remuneration under either the 1978 (and earlier) Act or the 1991 version which introduces 'dependence'. Under the earlier versions, a variety which is bred from a protected variety is not infringing (owes no royalties) as long as the new variety is distinct according to the UPOV interpretation. If the protected variety is a landrace which is used (as is permitted under the research exemption) in a breeding programme – a general case because landraces seldom are acceptable for commercial-type farming operations – the resultant new variety or varieties would get the sales with no payments owing to the owner of the landrace.

The 1991 UPOV Act rectifies that situation in part by differentiating between initial and essentially derived varieties, with essentially derived varieties requiring permission from the owner of the protected variety for marketing. In most cases that permission would be granted for a royalty fee. However, UPOV Article 14(5) establishes two conditions for derived varieties, that they be 'predominately derived ... while retaining the expression of the essential characteristics'. As an example, the essential characteristic could be disease resistance found in a landrace. In the breeding process, the remaining (undesirable) genetic material would be bred out so that the genetic composition of the resulting commercial variety would be predominately from another source. That would seem to preclude its being established as an initial variety under the proposed interpretations. Those interpretations also specify the existence of a single initial variety for any derived variety (UPOV, 1992). The interpretations are advisory only, and eventual national applications could be more favourable to PBR use for landraces. This is something for national governments to consider.

Trade secrets

Trade secret legislation allows those whose industrial secrets have been improperly acquired to use the courts to stop further use and/or seek restitution. They would apply if an employee changed employment to that of a competitor, there revealing the production secrets of the former employer. Or they would apply to outright theft, in general potentially any case in which: (i) the item or information had value and (ii) an effort was made to keep it secret. The community aspect of much cooperative technology makes secrecy problematic, and indeed secrecy would be contrary to the open exchange considered necessary for maximizing advances within agriculture. Thus trade secret legislation is not always applicable, but note should be made that secrecy protects many exchanged genetic materials, for example, under the Merck/INBio agreement (see Chapter 2). In such cases trade secret legislation would come into effect only if the secrets were improperly accessed.

Overall then, traditional forms of IPR are not really applicable to the major forms of cooperative technologies; certainly critics of IPR are correct in that regard (UNDP, 1994). Attention is directed next to alternative forms of IPR referred to here as 'non-traditional'.

Alternative forms of IPR

Intellectual property rights, as is suggested above, are but one means (and not a very applicable means) of claiming control of and remuneration from indigenous knowledge. Other possible approaches to be considered here include 'Farmers' Rights', treatments of folklore, codes of conduct, and appellations of origin. For a broader group, see Posey (1994).

Farmers' Rights

Farmers' Rights is the term developed by the FAO under the so-called Revised Undertaking for Plant Genetic Resources (see Chapters 2 and 6). While not necessarily restricted to plants with agricultural applications, it is quite evident that is the intended focus of the Undertaking. In Resolution 5/89 Farmers' Rights are defined as 'rights arising from the past, present and future contributions of farmers in conserving, improving and making available plant genetic resources ...'. Farmers' Rights are to be 'implemented through an international fund on plant genetic resources which will support plant genetic conservation and utilization programmes, particularly, but not exclusively, in the developing countries' (FAO Resolution 3/91, Annex 3 to the International Undertaking). No further details on the implementation and operation of this fund are included.

In concept, Farmers' Rights operate more as a moral obligation than an economic incentive; Brush (1996a, p.12) refers to them as 'retrospective equity.' They are not connected with any specific future action but rather with a general conservation and equity objective. The situation is described thus for the CBD (UNEP/CBD/IC/2/13, 1994, Par. 34):

Farmers' Rights could:

- ensure that farmers, farming communities and their countries, receive a just share of the benefits derived from plant genetic resources ... and thereby
- provide incentives and means for the conservation and further development of these PGR.

But nothing is included on the mechanism by which the payments provide an incentive for specific future actions.

Thus Farmers' Rights, it is noted without prejudice but only to emphasize that the objectives, and hence the likely results, are quite different from IPR. However, one parallel which has been drawn on several occasions (e.g. UNDP, 1994) is to blank recording tapes and other selected applications. There the very reasonable presumption is drawn that individuals will make copies, denying authors and artists royalties. The fund compensates those losses on some formalized basis; presumably the nationality and residence of the recipient would make no difference. A similar approach could be used for seeds and other genetic resources if a clear connection can be made between the injured parties and beneficiaries.

Perhaps the major comment which can be made is the lack of action on the fund since its proposal. The time span has been relatively short, but there are few indications to date that such a fund will be constructed, at least under these specific auspices. The entire International Undertaking process received much negative attention in the developed countries early on due to the interpretation of 'plant genetic resources' to refer to both un-improved and improved genetic materials (Article 5) (see Grossman, 1988). Private firms have not made their products available without charge, and while it is a matter of interpretation if that was specifically required by the Undertaking, it did poison the atmosphere. Subsequently, the proposed tax on seed sales was never supported.

It is, of course, possible that the concept of Farmers' Rights could be pursued more readily under a different name and institutional structure, possibly the Biodiversity Convention. For the present, implementation may come in India where one proposal for PBR legislation calls for Farmers' Rights (Swaminathan and Hoon, 1994, Articles 8, 9, and 23). However, because of the national form of that proposed language (as with all IPR legislation), the India-only system is effectively a national 5% gross seed tax with the funds to be returned to rural and tribal families contributing genes where identifiable, otherwise to a Community Gene Fund. The use of sim-ilar national legislation may be indicative of future steps; according to a working group of the FAO Commission (see Chapter 6), the lack of any agreed-upon definition of Farmers' Rights – and the fact that the CBD has shifted its focus from the 'common heritage of mankind' concept to the benefit-sharing and sovereign rights of states – means any real resolution will probably involve both national legislation and an international approach (*Diversity*, 1996a).

Folklore

Many of the issues associated with protecting genetic materials have paral-lels in protecting expressions of folklore. That is particularly true of land-races which, like folkloric expressions, are the result of long-term community contributions. And again like landraces, there is no system of compensat-ing, or even acknowledging, those communities for their contributions. The applicable IPR systems, copyright and trademark, operate similarly to patents in requiring new and unique creations, which folklore is not. Perhaps then attempts to protect folklore will provide some insights for use with genetic materials.

Treatments of IPR for folklore culminated in the joint 1985 'Model Provisions for National Laws' by WIPO and UNESCO (WIPO, 1985). There, the expressions of folklore are defined as 'characteristic elements of the traditional artistic heritage developed and maintained by a community

... or by individuals reflecting the traditional artistic expectations of such a community'. These expressions may be verbal (folk tales), musical or action (dances) as well as tangible expressions like art, musical instruments, and architectural forms (Model Law, Section 2). When used 'with gainful intent outside their traditional or customary context' such expressions are 'subject to authorization' by the competent authority of the community (Section 3). The expressions may originate from the community or elsewhere, provided they were subsequently further developed, adopted, or maintained through generations (Par. 35).

As can readily be appreciated, the issues are indeed similar to those for selected cooperative technologies like landraces. However, no helpful detail is included on how to implement what can only be described as concepts. For example, in the frequent situation where neighbouring communities practice slight variants of the same tradition, whose permission would be required, any one of the communities, or some/all of them? How or who would determine when an expression is different enough to be a separate form of expression? What competent authorities would be identified to represent a community? And what constitutes an 'artistic heritage'? Hence, the protection of folklore has moved little beyond the conceptual stage, and not far beyond probable L/I knowledge.

Codes of conduct

Codes of conduct refer to standardized but voluntary agreements specifying obligations. They are similar to a one-sided contract voluntarily entered. The FAO has over several years prepared a 'Code of Conduct for Plant Germplasm Collecting and Transfer,' still in draft form, which could serve as a model for protecting some cooperative technologies (FAO, 1993). Another Code is in use by Kew Gardens while the US National Cancer Institute utilizes a 'Letter of Collection' (reproduced in OECD, 1996, Annex V). Downes *et al.* (1993) have developed an elaborate code with elements for governments, local communities, collectors, and users.

The FAO Code, which is directed primarily to governments, has the principal objectives of promoting respect for the environment and local traditions and cultures, and establishing mechanisms for compensating local communities and farmers for their conservation and development activities (Article 1). The mechanism for achieving these goals is to require collection permits (Article 8) subject to certain conditions, including 'financial obligations', restrictions placed on the distribution or use of the germplasm or improved materials derived from it, the use of care in the collection process, and provision on request to the country of duplicate sets of the collected materials (Articles 8, 10 and 11).

Separate obligations apply to sponsors ('see to degree possible collectors abide by Code,' Article 12), curators (provision of further samples, Article 13) and users ('consider providing some form of compensation',

Article 14). This Code is seen as serving temporarily until national legislation is passed, or possibly a legally binding international agreement like a protocol under the Biodiversity Convention is reached. For the present, the Code can be seen in part as a model law for national governments (see Chapter 2). In its present form as a voluntary guideline, it has limited utility for protecting indigenous knowledge.

Appellations of Origin

Appellations of Origin are coordinated by the Lisbon Agreement of 1958, which, with its 17 members, is administered by WIPO. The Agreement (Article 2) defines applications of origin as the 'geographical name of a country, region or locality, which serves to designate a product originating therein, the quality and characteristics of which are due exclusively or essentially to the geographical environment, including natural and human factors'. The prototypical example is champagne from the region of the same name in northeastern France. Some national laws (e.g. in Brazil) also refer to geographical indicators distinct from Appellations of Origin, but as a practical matter they function similarly (see UNEP/CBD/COP/3/22, 1996, Par. 51(d)).

Extensions of this IPR approach to L/I knowledge are as yet untested. But the definition implies a quasi finished product which can be identified and distinguished by source. From that perspective, the concept would not seem to apply well to many L/I technologies. Yet for selected L/I technology contributions to such products as cosmetics, which use a combination of natural products for the overall effect (as opposed to pharmaceutical products where the causal agents must be characterized in detail), appellations of origin may be applicable. Hence there may be some scope of appellations of origin for L/I knowledge, but it would require a detailed and geographically dispersed knowledge of the plants involved.

In summary, existing forms of IPR, both 'traditional' and 'non-traditional', are not well suited to the protection of L/I knowledge. Similar conclusions have been reached by others using a variety of perspectives. Reasons are specific to each form of protection, but there *may* be more similarities among them than is immediately apparent. Indeed, sometimes two or more forms are utilized together as a technology may be patented with the best means of use protected by trade secrets or hybrid pure lines may be protected by both PBR and secrecy in a complementary manner. IPR of the types described here have two principal attributes: (i) a clear identification of what is being protected (needed to identify the limits of the protection); and (ii) a particular use or market (fulfilling the purpose of promoting the production of useful products as well as avoiding a clutter and research disincentive of protected products with no identified use). L/I knowledge may sometimes satisfy (i), but (ii) is often problematic. Thus existing forms are rarely useful and alternative approaches are needed.

Proposed New *Sui Generis* Systems

From the preceding it is evident that existing systems are not suited for the protection of indigenous knowledge, at least in general, although they may apply in some cases. Secrecy has applications for the communities themselves, but imposes costs on the remainder of society which may be denied useful products that would possibly be available on mutually agreed terms under other systems.

The general unsuitability is not surprising for two reasons: IPR was not designed for the protection of traditional knowledge. IPR systems, in general, are designed around specific objectives and are not readily extended for additional purposes.

Traditional knowledge is a more complex matter for IPR-type protection because the subject matter is more ephemeral. That is, while existing IPR protects knowledge embodied in some identified form, traditional knowledge is intangible. Here, an effort is made to identify a functioning system, a *sui generis* system. We begin with an overview of efforts to date, and then present an approach proposed by the author.

Prior efforts

To date, limited progress has been made on the development of a functional system. The Oxford Centre for the Environment, Ethics and Society (1996) has been the source of a proposed Traditional Resource Rights system, but one lacking 'practical instruments' at this time. As described, it is a 'bundle of rights', 'covered under a significant number of international agreements that can be used to form the basis for a *sui generis* system'. Yet the form of the *sui generis* system itself remains to be identified beyond the creation of an *'Ombudsman's Office'* to advise and represent indigenous groups. The proposed Traditional Resource Rights system seems as much a process as a goal at present.

One earlier proposal (Gadgil and Devasia, 1995) has been to enforce indigenous knowledge rights by requiring that the owners of the source materials be identified in any resultant patents. This would necessitate agreement on any subsequent licensing arrangement. In some instances, sources of materials are presently identified on patent applications so that what is proposed is a generalization of that practice. There are several reasons why this approach alone would not suffice for protecting indigenous knowledge (see OECD, 1996). Cited would be the owners of the materials, but the list may not include the owners of the knowledge. More generally, many products are never patented and if they are the patent may apply to a derivative compound suggested by the natural material, but not that specific material. This means ownership is unclear. Finally, a requirement for identifying source materials would require legal changes under many national

and regional IPR systems in the recipient (user) countries. That at minimum would be a lengthy process. What is proposed below is simpler and
more direct, although it would not preclude a naming in a patent when
appropriate.

Swanson (1995) undertook an ambitious comparison of the role of IPR
and 'knowledge derived from the biological activities of natural organisms'.
In simple terms, he argues IPR protects informational investments (human
capital) so that a parallel 'informational resource right' could be established
to protect 'natural resource investments'. In the former the industrialized
countries are well endowed, the latter the developing economies, so that there
would be some rough parity. What the proposal overlooks in this comparative analysis is the requirement that patents protect only knowledge/
information reduced to practice, i.e. functional inventions, not disembodied
ideas (the 'utility' or 'industrial application' requirement for patents; see
Chapter 9 for a discussion of the operation of IPR systems). Few natural
organisms are anywhere close to marketable products. Protecting a 'natural
organism' therefore has no direct comparability with existing IPR systems
so that Swanson's initial 30 pages of analysis are often misdirected.

Operationally, Swanson's system would function through materials
collected from 'biodiversity reserves', the applicable form of natural
resource investment for this system. 'A state's programme would qualify for
inclusion within the regime by means of investing in biodiversity reserves
and establishing prospecting programmes'. Discoveries there from 'should
be made subject to internationally-recognized exclusive rights on registration
with some sort of centralized office (analogous to a patent office)'. This
registration office would determine the scope of the right based on the
attributes of the genetic resource as well as the value and investment in
the reserve from which it was collected. Other key administrative matters
like duration are left indeterminate. Clearly, these ideas do not apply directly to indigenous knowledge and anyway are far from an operational system. Indeed, it would seem far simpler and faster to establish exploration
MTAs for the proposed biodiversity reserves stipulating that a commercial
use agreement must be established if a marketable product is identified
(see Chapter 2).

Stephenson (1994) similarly draws on an existing system as a model –
in his case, computer software licences. Sometimes referred to as 'shrink
wrap licences', the text may specify that opening the package indicates
an agreement to comply with the listed terms. Those terms are non-
transferability and non-exclusivity, sometimes involving an annual fee.
What is most relevant to Stephenson is the broad similarities with traditional
knowledge, especially its evolutionary nature (computer programs are frequently modified by suppliers and/or users). Yet the licence is for a program in a fixed form (e.g. Word Perfect version 6.1a) with possible updates
included. This is quite different from indigenous knowledge embodied in a

form/product lacking in specific interest to Western markets so the analogy is not persuasive. Consider, for example, that many traditional societies were known to cover cuts with soil or other materials. That hints at the existence of antibacterial agents, but is a far cry from penicillin. More relevant is the 'surcharge' collected on recording tapes in the USA, the funds used to compensate performers on the assumption that the tapes will be used for home recordings without paying a royalty fee. While the concept is useful, indigenous knowledge lacks a universal media, like recording tapes, to which a fee can be attached.

Butler and Pistorius (1996) draw a parallel between revealing indigenous knowledge and an earlier Dutch PBR-like law (1941–1966) which functioned without property rights. Under that system, farmers paid a levy based on the acreage of a crop, the funds being used to support plant breeding; farmers' subsequent use of varieties was unrestricted. Basing the tax on crop acreage is plausible (if providing no small data collection problem in a large, diverse agricultural system). But in the subsequent section Butler and Pistorius talk in terms of levies on individual varieties, which would seem to be impossible to monitor on a farm-by-farm basis. Conceptually, in their view, a similar levy approach could be used for rewarding L/I knowledge, provided an identifiable, observable, and unambiguous product could be found to which the tax could be attached. If that product were 'improved' seed in general, with the recipients of the funds local farmer-breeders, the Butler and Pistorius system is one and the same with some proposals for Farmers' Rights, taxing of improved varieties. But it would differ from the Dutch system where the variety development financed by the tax was of direct use by, and the research subject to some control of, the taxed farmers. These conditions do not necessarily apply in Farmers' Rights, so the idea requires further development.

Gollin (1993, p. 181) refers to a system of 'discoverer's rights'. These rights are operationalized through access legislation with the fee passed from the country 'along to the person or group that discovered or traditionally used the species'. That is, the proposal is for access legislation with sharing at the community level. The difficulty with that approach is the uncertain rights of many local and indigenous communities to genetic materials. Indeed, the focus on indigenous rights under the Biodiversity Convention can be seen in part as an effort to achieve greater use rights to traditional lands from national governments. Without that recognition, it is difficult to consider a genetic resources property rights system based on ownership of the materials themselves.

For that reason, I have previously (Lesser, 1994b) advocated a system which separates the ownership of the genetic resources from the ownership of the knowledge of its use, something undisputedly community property. The complexity created is how to devise a workable system which protects just the disembodied knowledge without unduly hindering the sharing of information. I prefer to refer to my proposed system as *reserved rights*, reflecting the situation that particular components of indigenous knowledge show promise but may not culminate in commercializable products (see below). Operationally, my proposed solution extends Gollin's (and others') suggestion to develop a rights scheme based on access legislation.

Proposed sui generis *system*

Principles

The proposed system incorporates several of the broad principles of standard IPR systems. Thus the general intent is to provide, to the degree possible, holders of indigenous knowledge with the same opportunities for controlling and profiting from their inventions as the developers of other classes of inventions. Equity in this form is the stated objective of many related efforts, such as Swanson's. The operating principles are as follows:

1. Developers of 'traditional knowledge' inventions have the right to control the use of their inventions, including benefiting financially from it.
2. All payments come from market sales rather than from contributions or other third party sources. Other forms of compensation may be appropriate as well, but are not part of this plan.
3. Recognizing the importance of the open exchange of knowledge for the general welfare, the system shall encourage revealing knowledge rather than maintaining secrecy. Indigenous knowledge holders, of course, maintain the right to keep knowledge private, but under the proposed system they cannot both maintain secrecy *and* participate in the proposed reserved rights system.

Conceptualization of operation

The preceding describes needs for protection of indigenous inventions with two principal attributes:

1. Traditional uses of materials often indicate some likely merit (e.g. activity for medicinals or disease resistance in a landrace), but typically no commercializable products have been identified. This lack of industrial use makes it both procedurally barred and financially impractical to attempt to patent a wide range of materials, the vast bulk of which will never have widespread commercial use.
2. Revealing the methods of use without some protection could mean loss of the knowledge; there would be no legal or practical impediment to its appropriation by other parties. Hence what is required is a means of revealing the knowledge while reserving the right to control its use.

The focus of this proposed approach consequently involves a *registry of traditional uses* of genetic materials, including a description of the genetic materials in sufficient detail so that they may be identified unambiguously. The registry may constitute multiple components including previously published scientific reports of genetic materials with use references (e.g. Levingston and Zamora, 1983) or current community-based catalogues such as those begun in India (Bhatia and Kothari, 1996) and the Andes (UNOCIAE-C, 1994). References pre-dating this reserved rights system may be allowed. Operational decisions on priorities of references and duration of validity of registry items along with other technical matters are discussed below.

The system still lacks an enforcement mechanism. Here the proposal is to embed enforcement within national access legislation. Prototype legislation is from the Philippines which specifies that a separate commercialization agreement must be negotiated if a market use is identified; without that agreement commercial use is prohibited (see also Downes *et al.*, 1993). Andean Pact legislation also requires that an agreement be secured first with the local community/private owner and subsequently with the State (see Chapter 3). Providing an enforcement mechanism for the Reserved Rights system would necessitate that national access legislation require a commercial use agreement be negotiated with the owners of the indigenous knowledge as well as with the State. Failure to comply could jeopardize future access rights to the country as well as raise questions about the validity of any subsequently acquired IPR as the result of a violation of the Convention's requirements for 'prior informed consent' and 'equitable sharing'. 'A company that ignores the CBD risks investing hundreds of millions of dollars in developing a new product only to find that it has no proprietary protection' (Belson, 1996). The filing of a name or names of individuals to represent the communities in negotiations would satisfy the 'prior informed' aspect of the legislation.

Operationally, inclusion of a requirement in national access legislation for negotiating terms of commercial use should be sufficient to protect the rights of local and indigenous communities. Practically, the process of prospecting and commercialization is facilitated because the prospectors through the 'registry of traditional uses' will have prior knowledge of the requirement for an agreement, and the party or parties involved. This process is significantly more transparent than that in the Philippines law, requiring the prospector to act 'in accordance with the customary laws'.

Politically, serious negotiations for the inclusion of indigenous rights in access legislation will be required in some countries. Among other matters, the addition of a stakeholder will be seen to reduce any compensation for the government; the market value of the product is unaffected, all that changes is the division of the proceeds. That perspective is unfortunate as indigenous knowledge can enhance the value of genetic resources by acting like an initial screening for activity. That has been the basis for the strategy employed by Shaman Pharmaceuticals (King, 1994; see also Asebey and

Kempenaar, 1995) as well as the Body Shop, a producer of cosmetics and personal care products. Lesser and Krattiger (1994b) provide an example of how the value of the indigenous knowledge can be computed. If the 'hit rate' from random samples is 1 in 12,000 which can be doubled to 1 in 6000 by using indigenous knowledge and a screen costs $100 each (numbers are illustrative only), then 'savings' from / value of this indigenous knowledge is minimally 6000 × $100 = $600,000 (royalty income would be unaffected). In this and other ways, effective negotiation processes should be able to separate the value of the knowledge from that of the materials themselves so that all parties benefit to the degree of their individual contributions.

Local communities furthermore can agree to withhold any objections to the commercialization of what they see as their knowledge. Generally, private companies will not become involved in products for which use rights are unclear. Hence a protest by a local community, possibly leading to a widespread boycott, would deter many firms from prospecting in a country. In that context, an agreement not to protest would have economic value for the government.

To this point in our discussion, the implications are that the system will be wholly internal to a country. The complication there is that knowledge may be acquired from one country and samples from another; the previously mentioned neem is an example (pp. 115 and 117). With only domestic legislation, the suppliers of the knowledge would likely not be compensated as collectors could simply identify any other country for prospecting; multiple examples exist of genetic materials being available in several countries. Two possible approaches could remedy this matter. First, communities may agree to register jointly (e.g. list multiple owners to a particular indigenous invention). This would reduce the opportunity for collection outside the reserved countries. It would also reduce competition on licensing terms among communities. Some prior agreements would be needed on who will negotiate for all the involved communities and how any payments would be divided. A second (and complementary) approach would be for national access legislation, on a reciprocal basis, to include registration lists from other countries. That is, countries A and B could jointly refer to each others' indigenous knowledge so that a prospector finding a valuable product in country B based on indigenous knowledge from country A would be required to compensate individuals in that latter country.

Sample legislation

The purpose here is the addition of further specificity in the proposal through the provision of sample legislative language. Each choice of language will be justified. Care should be exercised in using this text as it is intended primarily for illustrative purposes as opposed to forming the basis for actual legislation. Many potentially legally important details and definitions are excluded for the purpose of brevity and comprehensibility.

Clause for inclusion in access legislation. This research agreement provides rights for research purposes only. Any commercial use of the collected materials, or the compounds or structures identified directly or indirectly there from, is subject to a separate commercial use agreement. Mutually acceptable agreements must be negotiated with:

(a) the owner(s) of the genetic resources as identified in this research agreement, and

(b) for genetic materials listed in the Registry of Traditional Uses *included in the Regulations for this legislation, the community representative(s) identified therein.*

The purpose of this clause is to provide a mechanism for the enforcement of rights to control the use of indigenous knowledge. This is done through the mechanism of national access legislation; anytime a material collected under a research agreement is to be commercialized, the party must consider the rights of local or indigenous communities. Those rights are determined by inclusion in the Registry of Traditional Uses provided as a regulation of the access legislation. Regulations, which may be amended by authorized individuals through prescribed means, allow more flexibility than inclusion in the text of the legislation itself.

Regulations. Registry: The Registry of Traditional Uses *shall consist of references to published listings of uses of genetic resources for agricultural, medicinal and other uses. Listings shall include:*

1. Descriptions of use(s), to include all materials incorporated and any preparatory practices. In case of multiple listings of the same genetic materials, the first published listing shall have priority.

2. Scientific names, descriptions and/or photos of genetic resources employed in the above uses in sufficient detail such that the materials can be identified by knowledgeable individuals.

3. Identification of community group representative(s) to be contacted for conducting commercialization negotiations. Inclusion of these names signifies that the identified communities have been previously informed of the possibility of commercial use.

4. Locally published, out of print or otherwise difficult-to-acquire listings shall be deposited in the National library and there made reasonably available, including a photostat service at prevailing government rates.

This regulation defines what constitutes the claim of rights to indigenous knowledge. Claims may be established only through publication, which encourages knowledge to be shared. Communities may choose not to reveal their knowledge (this system is voluntary), but risk loss if the knowledge is acquired in another way. In this regard there are direct parallels with the use of patents vs. trade secrets (see Chapters 7 and 9). Published materials must be complete to the extent of describing the indigenous uses, the materials in a way which can lead to their identification, and responsible individual(s) to be contacted. This is done for practical reasons, so that collectors can know what prior claims may apply to which materials and the communities can assert their claims.

The publications may be of several types, including existing scientific studies and local efforts supplemented by the required reference to contact individuals. Locally published materials must be made available in a known location so as to be accessible to collectors and other interested individuals. With no mention of geographic restrictions, the possibility exists of listing uses in many countries. This would limit the opportunity for collectors to derive knowledge in one country while sampling in a second, with no legal obligations to compensate the indigenous knowledge.

The granting of priority to the oldest referenced use removes ambiguity and the spectre of unwieldy multiple claimants. The possibility of losing knowledge compensation to another claimant encourages cooperation in listings; a single listing could represent multiple communities, each identified as a contact name in the directory. For practicality, however, the groups must select a single chief negotiator to represent them in contract discussions. The division of any proceeds would best be determined prior to the preparation of the listing in the Registry.

Duration. Community claims to knowledge shall be limited to 50 years from first publication in the Registry of Traditional Uses, after which time the knowledge shall be considered to be part of the public domain.

The selection of 50 years is admittedly arbitrary; other periods could be substituted. What is important is the concept that rights have a temporal limit. The proposed 50-year period describes a time frame over which much indigenous knowledge will have been subjected to scientific testing, placing the issue of subsequent protection and compensation in a different dimension. The longer duration of protection than that of patents and PBR (about 20 years) can be justified because no actual products exist. The duration of any contractual agreement will be determined by the parties, but the duration of patent protection, in the cases where that applies, will certainly affect any agreement over contract terms.

Dispute settlement. Claims of indigenous use shall be subject to documentation and verification at the request and expense of the collector. Disputes shall be heard in the [nationally named] court. Parties may select to have disputes settled by binding arbitration.

This clause establishes the rule of law on claims of knowledge, while placing the financial burden on the collector (the disputing party). In line with other IPR disputes, the national court identified to handle IPR issues shall also have precedence in this area. Under GATT/TRIPs (Section III), countries are required to 'make available ... civil judicial procedures concerning the enforcement of any intellectual property rights ...'. However, binding arbitration is allowed as a speedier and less costly alternative. Multiple arbitration services presently exist, such as the World Intellectual Property Organization Arbitration Center (1993). Other available systems are identified in Wetter and Priem (1991).

Claims of indigenous and local use are subject to judicial review, or binding arbitration.

This clause makes it clear that the judicial review process extends to the verification of claims of indigenous use. It is not sufficient to state use; it must on demand be verifiable. Observation of use or sworn testimony should suffice in most instances.

Conclusions and Approaches

Those who complain of a double standard regarding the IPR protection of genetic materials have a legitimate position (e.g. ACTS, 1993; Greaves, 1994; UNDP, 1994). Conceptually, IPR protection for L/I technologies in the areas of genetic resources and landraces fits smoothly within the historical development of IPR legislation. Practically speaking, however, current legislation, for example, is applicable to improved plant varieties but is not really suited to landraces and the like, even though they are technically protectable. The closest current law comes to effective protection is the 1991 UPOV text. However, to be useful, a quite different interpretation (possibly involving a textual change) of a derived variety's attributes would be required. That option should be considered, but it could be a difficult and lengthy process. For other materials, patents are usable, but the costs of documenting and preparing an application make patents a prohibitive approach for the great bulk of materials of uncertain use and value, even if some funding system were established. Overall then, traditional IPR is not broadly useful for applications of L/I knowledge.

When the assessment is enlarged to include 'non-traditional' forms of IPR, Farmers' Rights and folklore, while interesting concepts, are in their present forms not really fully developed for the protection of cooperative technologies. Appellations of origin have promise in some product areas, such as cosmetics, but would take creative adaptation, and would be limited at minimum. The FAO Code of Conduct is less a form of IPR and more a model contract.

Clearly, if IPR is not widely applicable to L/I knowledge, then it is ineffectual in aiding either the creation/preservation of that knowledge, or its transfer, as specified in the CBD. Work to date has largely been focused on the definition of the problem. Now that there is broad agreement on the inadequacy of existing IPR, it is time, past time, to proceed to possible solutions.

A first step would seem to be the identification and resolution of obvious contradictions. As Gupta (1993) writes:

> On the one hand, many environmental activists the world over oppose modification in the patent regime in developing countries ... On the other hand, the same activists also wish that companies and western institutions which invest billions of dollars in research and product development should provide their technology without any cost to the developing countries.

One or the other – IPR everywhere or IPR nowhere – is the only reasonable course, and I am betting on the former.

Solutions for L/I knowledge would seem to come in the form of contracts (Material Transfer Agreements, see Chapter 2) and a *sui generis* system. The complexity of that undertaking should not be underestimated, as indicated by the inadequacy of proposals to date, for it involves protecting unembodied knowledge. That is at the same time quite different from the objective of existing IPR – meaning little experience is available to draw on – and risks confusing ownership and use by providing overly broad protection. Those concerns notwithstanding, I have attempted here an initial system explanation. The proposal is somewhat novel in operating as a reservation system based on a registry of uses. Further developments and assessments are required.

Certainly, the private sector needs to be more involved in these discussions than has been the case in past CBD deliberations. While firms are understandably reluctant to enter an environment often characterized by antagonism, representatives are none the less to be faulted by not understanding the significance of the CBD and effectively relinquishing the entire educational role to NGOs. Some NGOs, in my view, take an extreme view which is not useful to the future of biodiversity, or the CBD. Sing Nijar and Yoke Ling (1994) of the Third World Network, for example, come out strongly against patents, 'There should be no patenting of any form of life ...'. That is not a meaningful position in the present context (Brush, 1996a, p. 11), nor would the complete revocation of IPR achieve the new world order which seems to underlie much of the Third World Network rhetoric. Others seem to bait the private sector in ways making an accommodation more difficult. Greaves (1994), in an example he kindly footnotes as imaginary, develops a scenario around 'Voracious Pharmaceuticals, Inc.'. That is silly and counterproductive.

Rather, there needs to be an accommodation among the users and suppliers of genetic resources, replacing the current climate of mutual mistrust. This does not suggest the sides will fully accept the perspectives and motivations of the other; at some level buyers and sellers are natural antagonists. Rather it involves a recognition that there is mutual benefit to be achieved through careful, structured cooperation on mutual consent. Legal systems such as IPR law provide the structures for such voluntary collaboration. That is why serious work is so urgently needed in the area of IPR for L/I knowledge. That though will take some time.

More immediately, there have been proposals for some kind of international fund and/or ombudsman role to assist with application costs for IPR (UNDP, 1994); Gupta is attempting to implement the approach for India. That will not resolve the underlying issue – the inapplicability of IPR to most forms of L/I knowledge, but it is a start. Public and private sources, including the private sector, should give a high priority to funding that system.

Chapter 8

National Technology Transfer Commitments Under Article 16[1]

The objectives of the Convention on Biological Diversity, as described in Article 1, are as follows (emphasis added):

> the conservation of biological diversity, the sustainable use of its components and the fair and equitable sharing of the benefits arising out of the utilization of genetic resources, including by appropriate access to genetic resources and by *appropriate transfer of relevant technologies* ...

While the sentence structure leaves this statement open to several interpretations, the general sense suggests that access to genetic resources and technology transfer are *one* mechanism for achieving the goals of conservation and sustainable use. Moreover, because transfer is the only mechanism specifically identified, it may be inferred that technology transfer is to be accorded special significance. The Convention does not attempt to define 'technology transfer', and 'technology' is characterized only so far as it 'includes biotechnology' (Article 2).

The Parties have, throughout the duration of the Convention, consistently placed special emphasis on technology transfer. Three decisions in COP1 related to technology transfer: I/7 called upon the Subsidiary Body for Scientific, Technical and Technological Advice (SBSTTA) to advise COP2 in this regard, I/9 placed technology transfer in the Medium-Term Programme of Work, and I/2 established it as one of the programme priorities. Also, II/4 supported a decision by the SBSTTA in its own Recommendation

---

[1] This chapter draws particularly on Lesser and Krattiger *et al.*, 1995.

I/4 to prepare a substantive background paper with recommendations. COP3 decided to continue the discussion at COP4.

Construed broadly, technology transfer incorporates not only the supply and access of the technology itself (Article 16(1)), but the terms of the transfer (Article 16(2) and (3)), the function of IPR (Article (16(2) and (5)), and the role of private sector (Article 16(4)). Additionally, there are matters of joint research and development (R&D) as well as capacity building and information exchange (Articles 12, 17, 18, and 19). These can be partitioned into tasks for providers (transfer, financing, private sector, IPR), tasks for recipients (capacity building), and joint responsibilities (R&D). Here, focus is on the roles of *technology* providers. That is done because there is some practical symmetry with the view that developing countries provide the genetic resources (which are a form of technology in their own right) while developed countries provide the technologies.

This partitioning should not be overstated, for technology is a worldwide product for which no area has a monopoly on supply, nor is there much prospect of a direct exchange of technology for germplasm. Such an EXCHANGE approach was proposed in the initial conception of Farmers' Rights (see Chapters 2, pp. 24–25, and 7, pp. 121–122) and failed completely, with lingering resentments. None the less, the Convention does make special reference to the roles of developed countries in technology transfer to developing countries (Article 16(3), emphasis added):

> Each Contracting Party shall take legislative, administrative or policy measures, as appropriate, with the aim that Contracting Parties, *in particular those that are developing countries*, which provide genetic resources are provided access to and transfer of technology ...

And again in Article 16(4), emphasis added:

> Each Contracting Party shall [ ] facilitate [ ] access to, joint development and transfer of technology [ ] for the benefit of both governmental institutions and the private sector *of developing countries ...*

Finally, the success of developing countries is made partly contingent on developed country actions (Article 20(4)):

> The extent to which developing country parties will effectively implement their commitments under the Convention will depend on the effective implementation by developed country parties of their commitments under the Convention related to financial resources and *transfer of technology ...*

The purpose of this chapter then is the assessment of the roles which developed countries might take in facilitating the transfer of relevant technologies to developing countries, and a cataloguing of commitments and activities undertaken to date. Considered first is the general relationship between the CBD and technology transfer, followed by a theoretical overview of technological transfer. The third section considers the more practical factors followed by an overview of government plans to date. The final section contains some specific suggestions on needs and approaches.

Relationships of Convention Objectives to Technology and Technology Transfer

The Biodiversity Convention appears to take a pragmatic approach to technology; technology is a, one, possible means of achieving stated objectives. Those objectives include conservation and the sustainable use of genetic resources. No distinction is made between technology which can contribute to conservation directly by, for example, providing an alternative source for a material once harvested from the wild, or indirectly by enhancing the efficiency of agriculture and hence reducing the need for clearing additional land.

The distinction is drawn with regard to sources of technology, as it is implied no country is self-sufficient in relevant technologies. Article 16(1), for example, makes reference to facilitating 'access for and transfer to other Contracting Parties of technologies …'. Moreover, developing countries are directly identified as potential recipients of technologies, 'transfer of technologies … for the benefit of both governmental institutions and the private sector of developing countries …' (Article 16(4)). These clauses recognize the current reality that technology is an international market and that developing countries as a group are net importers of technology.

Within this Convention framework then, technology is a means to achieve objectives, and technology transfer is a means to provide technologies where needed. It should be emphasized again that technology is but a method for achieving objectives, possibly a preferred one, but certainly not the only means. The complication arises that, while much technology is in the public domain, much is also held by private interests. Indeed, the bulk (up to three-quarters) of agricultural biotechnology research is estimated to be undertaken in the private sector, a departure from the previous round of publicly funded agricultural research (Persley, 1990). By 1997, 70 of 74 (95%) of commercial release applications for agbiotech products were from the private sector (James and Krattiger, 1996, Table 11). That means access is feasible only if the requirements of the private owners are satisfied. And since many of those technologies are patented or protected by other forms of intellectual property rights, they too must be respected. 'In the case of technology subject to patents and other intellectual property rights, such access and transfer shall be provided on terms which recognize and are consistent with the adequate and effective protection of intellectual property rights' (Article 16(2)). While interrelated, a full discussion of IPR is deferred until a following chapter (Chapter 9). Here emphasis is on governments' commitments and roles.

Within this conceptual framework, biotechnology is not distinct from other forms of (largely protected) technology and warrants no special attention. However, because it is the only form of technology identified specifically in the Convention, some additional comment is warranted. Biotechnology is distinct in several ways. First, the term is a shorthand means of saying '[bio]technological application' (Article 2) as, strictly speaking, biotechnology is a methodological approach, not a class of products. This distinction is important because the biotechnology process can lead to many and diverse products. As delineated by the Open Ended Intergovernmental Meeting of Experts, relevant products could be in the areas of environmental remediation, agriculture, industrial processes, and pharmaceuticals, among others (UNEP/CBD/IC/2/11, 1994; see also Yuthavong and Gibbons, 1994, and Ollinger and Pope, 1995). Furthermore, these products potentially can both protect biodiversity, directly through remediation and indirectly through enhancing agricultural efficiency, and create a market for genetic resources, as in pharmaceuticals and agriculture. Finally, at this early stage of the application of biotech products, it is not clear what the practical promise of this wide range of possibilities will prove to be. Because of these diverse yet uncertain roles, biotechnology receives special attention within the Convention. It is important to recall, however, that the references are to a set of quite distinct potential products, many of which will have to be treated very differently in the transfer process. There is no single 'biotechnology'.

Concepts of Technology and Technology Transfer

Anderson (1989) defines technology transfer as having occurred 'when a country acquires, imitates, or adopts technology developed elsewhere'. And the term technology itself has been referred to as applied knowledge with a problem-solving intent. UNCTAD (1990) employs a related definition, the 'transfer of systematic knowledge for the manufacture of a product, for the application of a process or for the rendering of a service'. Because problems occur in a particular physical, economic and social environment, technologies must frequently be adapted locally to function across broad areas. Hence, DeGregori (1985, Chap. 3) refers to technologies as being 'both universal and particular'.

Other definitional approaches distinguish between so called 'soft' and 'hard' technologies. Hard technologies incorporate the common concept of a machine, the embodiment of knowledge in a tangible form, while soft technologies are know-how, skills and techniques, the unembodied form (Glowka *et al.*, 1994, p. 85). From an economist's perspective, technologies are efficiency-enhancing by allowing the same level of production at lower levels of inputs (Peterson and Hayami, 1977). When the technological change uses the same proportion of inputs of land, labour, and capital, it is referred to as neutral. New technologies that change the so-called factor shares to (for example) proportionally more capital than labour would be called biased technological change. Finally, when efficiency enhancements can be traced to capital or land or labour, they are known as embodied. However, a shift with no clear causal factor is described as disembodied change. Such shifts can be attributed to enhancements in management or knowledge in general, themselves a form of technology.

Still a different perspective is that between the termed 'institutional' and 'cooperative' innovation systems. Institutional innovation refers in general to the scientific approach of specific and distinct objectives with a systematic path to a product, the whole often based on an individual's profit motive. Cooperative innovation for its part is a more tradition-based system with diverse objectives shared as to effort and benefit on a communal basis (UNDP, 1994; see also Chapter 7). In short, one applies to 'modern' methods while the other was and is associated with traditional peoples. The cooperative approach, because of its more diffuse objectives among other differences, would be expected to lead to fewer large, seemingly discrete changes commonly referred to as *inventions* to distinguish them from the slow evolutionary process of technological change.

These several approaches to the concepts of technology, technological change, and technology transfer reinforce the notion that this is both a complex and fundamental process. At the same time, the distinctions of usage should not obscure the fact that the basic concept of technology is similar across groups. Technology is adapted, applied knowledge, whatever the source and means of generating that knowledge. That is, there is, at this level, no real distinction between traditional and scientific knowledge. Technology transfer is simply the movement of technologies to additional applications; transfer may be geographic (the general concept) but could refer as well to a different product application in that same location. Where the distinctions become important is in identifying the ramifications of technology and technology transfer. Labour saving technologies, for example, are of social significance in areas with unemployment or low wages in general, but conceptually they are all technologies. Similarly, hard and soft technologies are conceptually the same, but the transfer process and the mechanisms to protect a soft innovation are often more limited, which leads to various controls over use, such as secrecy and patents.

Formal treatment of technology transfer

This conceptual introduction builds on the economic definition of technology from Peterson and Hayami (1977) where technology is, 'the phenomena of input quality improvements or an increase in knowledge leading to an increase in output per unit of input ...'. In simpler terms, this

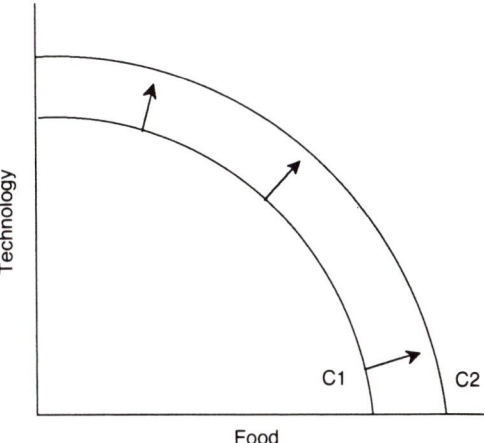

Fig. 8.1. Production possibility curves.

means more output for the same level of inputs, more product per labour hour, more food per hectare, etc.

In theoretical economic terms this can be represented as the outward shift of the Production Possibility Curve (PPC). Figure 8.1 shows a PPC for two products which indicates the maximum production of the two products (or each individually at the axes). The curved lines indicate that the production becomes less efficient as output is concentrated on one product or the other; without that effect the lines would be straight. While curved lines are more likely in practice, conceptually the situation is identical.

Technological change is represented by the outward shift from C1 to C2. For simplicity it is presented as a parallel change. Conceptually, the curve could be of any shape so long as C2 never touched or crossed C1. The compelling feature of technological change is that it provides more for less, more output for the same or less input. Technological change has permitted the general rise in the standard of living over the past centuries, food production being a case in point. Total agricultural production in 1994 would, using 1960s technology, have required about 40% more land than was employed. Much of the increased output, of course, is due to the substitution of other inputs, especially inorganic fertilizers. But the real technical change has been in improved varieties, such as shorter stature plants which have proportionally more grain overall. It is estimated that seed-based improvements have provided half the registered increase. In other areas such as electronics, technology has led to the production of new products, not only improved ways to produce existing products.

Theories of technological change

There are three principal theories of technological change. Many of these developed with specific reference to agriculture as one of the more fundamental and universal of technologies, but the concepts are fully general. The three theories are treadmill, induced innovations and diffusion (literature review in Kinnucan *et al.*, 1989).

The *Treadmill Theory* posits that the early adopters gain all the economic benefits from technological change. As technology by definition is output enhancing (for a given level of inputs), then total output rises with increased adoption. For the first entrants, profits are higher with stable prices and more saleable product with no comparable increase in inputs. However, eventually the increased output will cause prices to fall (demand curves are presumed to be downward sloping). This means that non-adopters are suddenly worse off with no change in input costs but lower market prices for their products. For them the choice is to adopt the new lower cost technology or go out of business. The process may be slow or rapid, but the outcome is the same: effective technologies eventually become ubiquitous.

While the treadmill theory may explain why technologies once developed are adopted, it says nothing about the direction of technological change. The *Induced Innovation* theory was first developed by Hicks and expanded by Ruttan (1984) among others. The theory states that inventive work is focused on changes in relative prices. Thus if the cost of labour rises relative to capital, labour saving technologies will be targeted. However, relative change is not the only causal factor as, in our example, labour saving technologies may be more difficult (and hence expensive) to create than capital saving ones. Capital saving technologies will then be targeted. Hence any change in input prices, even a general increase, can lead to the search for technological change. Alternatively, if labour is relatively expensive compared with capital, then a rise in labour costs would still prompt a search for capital saving technologies. In developing countries, the more likely scenario is high cost capital which explains why technologies are often labour saving in low labour-cost economies. The general point is that technological change is in the direction of the greatest cost saving. Once developed, technologies are useful only if employed, which is the subject of *Diffusion Theory* – who adopts what technologies and for what reasons. While some of the work in this area has been conceptual, focusing over time on sociocultural resistance and differential access to information (communication and education), much of the analysis has been empirical. Profitability has been perhaps the major factor in explaining relative adoption rates. Profitability has several aspects, the obvious net returns as well as access to capital (assets) and ability to bear risk. These concepts have proven especially useful in understanding barriers to diffusion (see Chapter 8, p. 155).

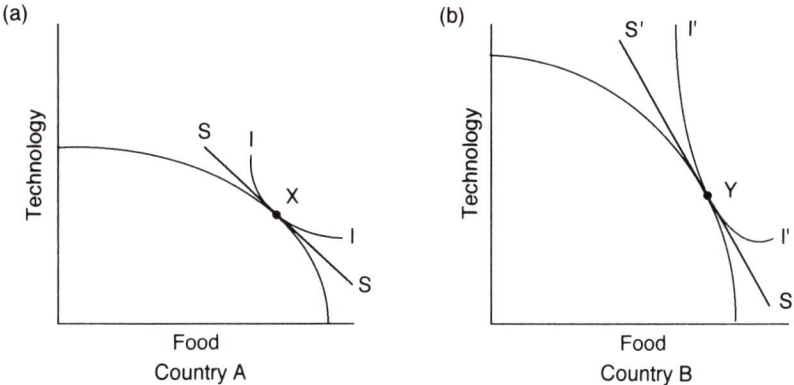

Fig. 8.2. Production possibility curves for countries A and B; case of absolute specialization.

Trade theories: comparative advantage

If, indeed, the preceding theories explain in part the development and diffusion of (or more correctly, demand for) new technologies, trade theories help explain why technologies are acquired from abroad as opposed to being locally developed. The best accepted theory is that of comparative advantage (see any trade theory text, e.g. Root, 1994). Comparative advantage demonstrates (here in the simplest two-country, two-product model) that trade can enhance the well being of both countries if they are relatively more efficient in producing one of the two goods. There are several subsets of this theory to be demonstrated before it can be accepted as universal. The major options are when: (i) each country is more efficient in producing one of the goods, and (ii) when one country is relatively efficient in both goods. Consider case (i) first.

Assume country A can produce food and technology in the proportions shown in Fig. 8.2a. This is similar to the production possibility curve (PPC) of Fig. 8.1. Without trade, A's citizens must be satisfied with some combination of products along this line, say point X. Technically, this is the point of tangency of the national indifference curve II for food and technology. A similar derivation can be used to place country B consumption at Y (Fig. 8.2b). Note, however, that the efficiency of production of the two goods varies between the two countries. Country A is more efficient at producing food as more than one unit of food must be sacrificed to produce a unit of technology. The reverse is true of country B. The efficiency of production is represented by the slope of the lines SS and S'S' respectively, known as the international commodity terms of trade (ITT).

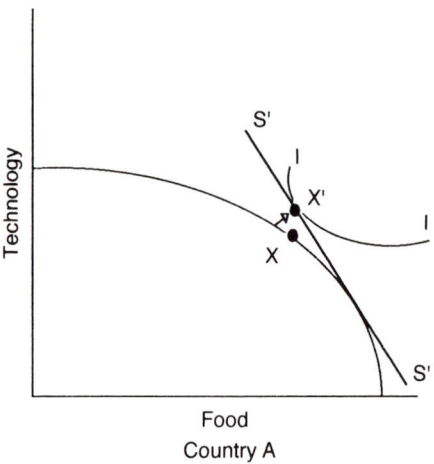

Fig. 8.3. Country A benefits from trade: case of absolute specialization.

Now consider the effects of trade. Draw the ITT for country B (S'S') tangent to the PPC for country A, as shown in Fig. 8.3. With the allowance of trade, the PPC becomes S'S' to which the indifference curve is tangent at X'. As X' allows for increased consumption of both food and technology compared with X, country A benefits from trade. Figure 8.4 shows the parallel situation for country B.

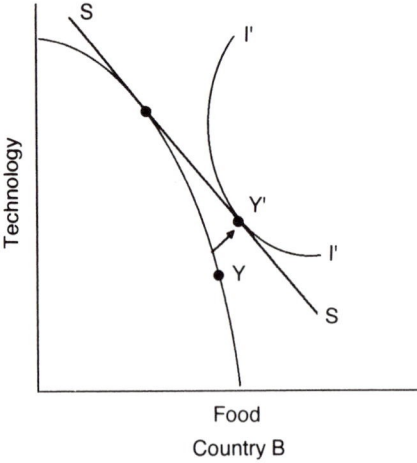

Fig. 8.4. Country B benefits from trade: case of absolute specialization.

Differences in factor endowments are one cause for the relative specialization between the two countries. To pick a real-world example, the US and Japan, or Zambia and Zimbabwe, are differentially endowed with land so that land extensive activities like grain production would be relatively less expensive in the US and Zambia. However, relative specialization is not necessary for the results. Differences in tastes will lead to the same outcome.

The case of one country (C) being absolutely efficient in the production of food and technology is shown in Fig. 8.5. This case might arise for example if one country was in a very harsh and expensive locale, say Iceland (country D). In Fig. 8.5, the PPC for country D is everywhere dominated by (lies inside of) country C's PPC. However, as long as the two PPCs are not parallel, that is, there is some relative specialization, then trade will benefit both countries. This is demonstrated graphically in Fig. 8.6 where the national indifference curves (II) are identical. This, of course, is why the theory is called comparative advantage. It is not necessary to have an absolute production advantage to benefit by trade, which significantly generalizes the theory. In the two goods case shown, the economic standard of living of country C will be higher than that of country D, C being more efficient. In practice, economies would produce thousands of goods so the comparison would not be so simple.

Regarding the actual operation of comparative advantage, Krueger (1983) has documented them in a series of detailed country studies. Rapley (1996, Chapter 4), however, concludes that comparative advantage benefits most relatively industrialized economics. Poorer economies lack the industrial base to compete in manufactured goods while the inelastic nature of demand for commodities means that prices fall rapidly in response to increased output.

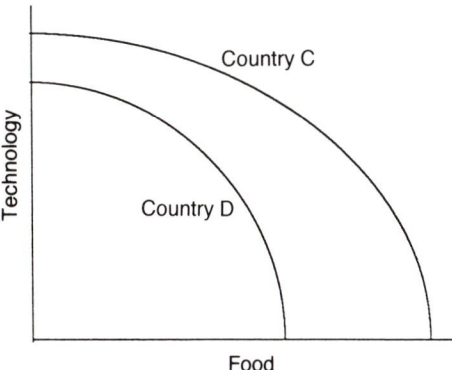

Fig. 8.5. Production possibility curves for countries C and D: case of relative specialization.

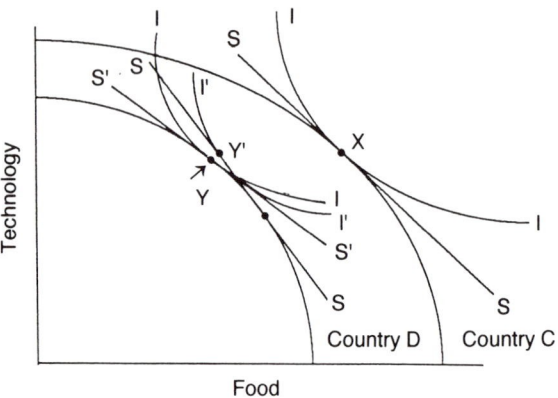

Fig. 8.6. Benefits from trade: case of relative specialization.

Attributes of Technology Trade

While the comparative advantage theory explains in a general sense the prominence of trade, it is completely general as to the type of products. For technologies, there are several reasons to expect they would be traded in disproportionate amounts. These factors are related to the nature of development of technology, and to the product's attributes.

The development of complex technologies tends to be education intense and, depending on the form, equipment intense as well. Given the uncertain nature of the results, technology creation tends to be risky. Taken together, these factors explain why some kinds of technology tend to be produced by larger private firms or by governments of wealthier countries. These two entities have the financial resources to absorb the uncertainty of research and development (R&D) investments.

To the degree technologies are knowledge based, they are non consumable goods. Non-consumption goods are those which are not diminished when used by an additional person. Sharing food always involves a reduced amount for the supplier, while an idea can be shared with no diminution. The capturable value of the idea may be diminished and is the reason for the use of patents and intellectual property rights (IPR, see Chapter 9). But the quantity of the idea is not reduced. For these two reasons, technologies are ideally suited to be traded, which presumably is why the concept is incorporated into the Biodiversity Convention. Indeed, world patent data show that, with the exceptions of the US and Japan, the great majority of patents in a country, 85% and more, are awarded to foreigners. This is another way of indicating that technologies are international.

Technologies, however, must typically be adapted to local conditions and needs. Thus trade is in the general, not the finished, form. Some adaptation is purely technical. But an equally important aspect is cultural. Because technology is ideas, it is inherently cultural, which both shapes and is shaped by technology.

Stages, methods and forms of technology transfer

Like many things, technology tends to follow set stages and forms. Understanding these typical patterns is helpful in understanding the process and ways to intervene when appropriate.

Firms entering international technology markets typically go through a two tiered decision process. First the choice is between exports of goods and production in the target country(ies). Second is the decision of the business form for local production, whether licence, joint venture or foreign affiliate (also known as foreign direct investment) (Vaughan *et al.*, 1994). Export decisions are based on a range of criteria, including national and recipient country conditions. Decision factors include market knowledge and size, risk, size economies in production, need for local adaptation, and recipient country regulations. Typically, firms prefer to enter new markets with exports before making the longer term, and more risky, commitment to local production (reviewed by Primo-Braga, 1995).

When exports are not possible or permitted, firms consider alternative forms of local production. Licensing has the benefit of affiliation with established local partners and less investment risk. Disadvantages are limited control over losing the technology and quality assurance (Barton *et al.*, 1989). Joint ventures are often mandated by national law stipulating majority national ownership. Joint ventures are riskier because they imply a fixed in-country investment. Subsidiary operation is, at the same time, the source of the greatest control and the riskiest due to the fixed investment. Foreign direct investment accounts for more than 60% of the flow of technologies to developing countries (UNEP/CBD/SBSTTA/2/6, 1996, Par. 39).

Associated with these methods of transfer is the forms. The most immediate is the transfer of products for direct use. These must be products of broad adaptability or larger markets justifying the expense of adaptation in the home operation. Capacity transfer refers to the dissemination of the capability for local production. This would be compatible with licensing, joint ventures, or direct investment. Finally, design capacity refers to training and related investments leading to the ability for local adaptation or further product development. Because design capacity carries the risk of creating a competitor if control is lost, subsidiaries are preferred for this type of transfer.

It is important when discussing technology transfer to remain cognizant of the appropriate form and stage. Many developing countries are interested in design capacity transfer as indicated in the wording of Article 18 of the Convention. Design capacity from the recipient country perspective translates into value added. Yet because this is often the most complex and risky activity for the supplier, opportunities will necessarily be more limited.

Overall, multiple systems and approaches for technology transfer already exist, as is appropriate for the wide range of technologies available (UNEP/CBD/SBSTTA/2/6, 1996, Par. 43). There is no reason to believe those mechanisms will not serve under the CBD as well as that no new systems appear necessary, a major benefit. But assistance will be required in making these systems operate in a timely and equitable manner for developing countries.

Factors encouraging adoption

Rogers (1962) developed the concept of relative advantage to describe the degree to which new technologies are preferred to old. He posits that preferences in the aggregate depend on sociological as well as economic factors. He identified four such factors:

1. *Compatibility*: the extent to which a new innovation is consistent with the existing norms, values and prior experiences of potential users. Also to be considered are the degrees of physical and managerial compatibility with existing practices.

2. *Complexity*: the extent to which new techniques and their consequences are difficult to understand. In general, less complex ideas are more widely and universally utilized (Graham, 1956; Kivlin, 1960).

3. *Divisibility*: the extent to which innovations can be used on a limited basis. If small-scale trials are possible, the risks are much less than for an all-or-nothing commitment.

4. *Communicability*: the relative ease with which knowledge of an innovation can be passed along. Communicability involves, in addition to the complexity of the innovation, the rapidity and tangibility of the benefits.

Biotechnologies, considered as a group, fall about midway in the four-dimensional adoption space structured by Rogers. Biotechnologies are relatively complex, and when the release of living organisms is involved, the biosafety risk is a significant technical effort to determine. Biotechnologies additionally do not always fit within the values of individuals and groups. On the other hand, biotechnologies can be highly divisible – within agriculture as little as a single plant – and may reduce rather than increase management requirements, as with disease-immune agricultural materials (but not those with disease resistance only). Biotechnologies, as well, tend to have rather immediate and tangible outcomes when commercialized.

From a governmental/policy perspective, most of these factors are controlled externally. Several, however, can be affected, indirectly influencing adoption. Certainly the communicability of technology attributes can be affected by government action. Additional attention can be directed to making innovations compatible with developing country users. This, too, is a possible area for assistance by the public sector.

Risk is a recurring theme with new technologies. While a number of risk-shifting approaches exist, such as the supplier retaining ownership and operating on a piece work basis with users, most are beyond the control of governments. Where governments can contribute is with divisibility. At one level, divisibility is a firm-level reflection of the localness of technologies. Each adopter wishes to know how an innovation will perform in his/her operation; nothing else really matters. Divisibility allows such a small-scale test. When that is not technically feasible, governments can assist with localized demonstrations allowing potential users to examine the effects under very specific conditions.

Government policy

More specific to the roles of government, Pray (1985) and Pray and Echeverria (1989) identify nine policy areas with significant potential impacts. De Janvry and Deither (1985) provide a more theoretical treatment of the same subject. Critically, it should be noted that many of these policies can be followed in supplying as well as recipient countries.

The nine policy areas, adapted for the purposes of this book, are as follows:

1. Regulation of new technologies.
2. Controlling the size of royalties and other terms of trade by affecting the terms of technology transfer.
3. Import/export restrictions.
4. Restrictions on foreign companies.
5. Intellectual property rights and enforcement.
6. Price controls.
7. Tax incentives.
8. Local technology capacity.
9. Promotion of appropriate technologies.

Of these, several require particular attention. Since profitability influences both the demand and supply of technology, anything like price controls or restrictions on foreign firms or royalty rates has a direct bearing on technology transfer. Many profitability-related factors are likely to be controlled by the recipient developing countries.

At a country-specific level, recent experiences in Brazil are revealing (*Economist*, 1996a). There, mergers and acquisitions shot up 50% from 1994 to 1996. Underlying the increase is the decline in inflation and the tariff-reducing effects of Mercosur, the regional trade agreement. Simultaneously, these restricted the opportunity of the small local firms to raise prices while increasing international competitive pressure. Foreign firms saw a profit opportunity in transferring cost-reducing technologies and management practices, while the local owners recognized the need to become more competitive. Recent enhancements in Brazilian IPR, including a maximum 30-day registration of technology transfer agreements necessary to effect protection against infringement, also contributed to the willingness of companies to transfer state-of-the-art technologies (de Freitas and de Carvalho, 1996). More generally, there is a need 'to create policy and legal measures that form the foundation of technology partnerships' (UNEP/CBD/SBSTTA/2/6, 1996, Par. 43).

Supply country policy has ramifications as well. In the US, for example, firms are forbidden to offer bribes internally as well as externally – a commendable policy. The complexity arises over the definition of what constitutes bribery. In an extreme form it would be any payment to, say, a foreign government official; a milder definition with more cultural latitude would allow payments when intended to assist the agent in the assigned job performance (as a 'tip' for faster service), but not otherwise. Hence, internal policies on the external activities of national companies can influence technology transfer and should be considered in an international perspective. The complex subject of overseas earnings taxation also affects the firms' willingness to involve overseas technology transfer as well as the form and level of investment.

The regulation of technologies is often an expensive proposition and hence a major factor in influencing technology transfer. Certainly this has always been the case with living organisms, and that will not change as developing countries implement biosafety regulations. In the past, smaller markets/countries have had difficulty in accessing technologies when the national registration process was costly compared with the size of the perspective market (Heureux et al., 1992). For example, Cameroon charges firms for registration trials for pesticides. The multi-year trials cost FCFA 7.5–14 million (about US$30,000–56,000), or twice that considering intra-firm costs, per formulation, per dose, per pest, and per crop. These costs must be considered in relation to the banana and coffee pesticide markets of FCFA 1.5 and 0.1 million respectively in 1992 (Heureux et al., 1992).

But by far the most significant factor is the private sector. By one estimate, over 90% of technology transfer is in the form of commercial transactions among businesses (UNEP/CBD/SBSTTA/2/Inf.2, 1996). While that percentage would likely be lower for transactions with foreign countries, given the typically greater role of the government in the economy, particularly as regards agriculture, the general point would still apply. Technology transfer, in the narrow definition of the term, is predominately a private sector activity.

Private sector technology transfer is carried out in a number of ways, including exporting, licensing, joint ventures and wholly owned foreign enterprises. Whatever the form, firms must have some type of firm-specific advantages to compete successfully in foreign markets. Thus, decisions are firm-level. Entry may be horizontal – replication of the national business in other markets – or vertical, to take advantage of cost savings in other regions. With technologies, the motivating factor is often the exploitation of the R&D investment (Grubaugh, 1987). Numerous studies have evaluated the determining factors within specific sectors (e.g. Primo-Braga, 1995; Sheldon and Abbott, 1996). In general, those studies concur that governments effect the transfers largely through establishing the proper conditions for profitable investments.

Of course, not all useful and needed technologies are in the private domain. Indeed, Juma *et al.* (1994) note that 'most of the technologies presently needed by the developing countries are in the public domain'. The validity of that assertion exceeds purposes here. More relevant is the recognition that purchased technologies frequently include assistance on adaptation and management, as well as practical hints on efficient use. It is for those reasons that technologies closely related to public sector ones are none the less often purchased. Again, whether that constitutes a wise decision in a particular case exceeds purposes here. What is relevant is the human technology assessment capacity which would permit countries to assess the options and, if the choice were for the public sector technology, to provide for the necessary adaptation and management. It is for that reason there are widespread requests for capacity building (UNEP/CBD/SBSTTA/2/6, 1996, Par. 19–21), a traditional but none the less valid public sector role.

A third approach is the several public entities which, for the past few years, have been involved in facilitating transfer of biotech products in particular between developed and developing countries. This includes ISAAA, the project-specific bilateral United States Agency for International Development (USAID), Agricultural Biotechnology Support Programme (ABSP) coordinated by Michigan State University, the National Agricultural Research Systems (NARS), many of the CGIAR's 17 institutes, the technology-oriented CAMBIA in Australia, the methods-oriented Biotechnology Research Group of the Food and Agriculture Organization, and the policy-oriented IBSP of ISNAR, among others. All these institutions are involved in technology transfer on a non-profit basis.

National Plans and Programmes

As noted above, there is a general expectation under the Convention – the taking of 'legislative, administrative or policy measures' to facilitate access to and transfer of relevant technologies (Article 16 (2) and (4)) – for developed countries in particular to develop plans for assisting in technology transfer. The purpose of this subsection is the outlining of those plans to date, although, as with much of the Convention, little specific has been accomplished. That perhaps is not surprising, with much of the technology of interest owned by the private sector and outside the direct control of the government, but, none the less, planning to the present has been limited.

OECD

In 1995, the OECD distributed a questionnaire to its member states (the developed countries) regarding technology transfer, IPR and genetic resources under the Convention. Objectives related to 'current practices, experiences and expectations related to technology transfer in biotechnology' with particular attention to IPR (OECD, 1996). Technology transfer itself was construed broadly so as to include education and training through specific transfer mechanisms like licensing and joint ventures. While replies were eventually received from 19 countries, contributors within countries were limited so that the responses cannot be considered to be representative national views.

In general, the responses indicated that the transfers of relevant technologies have and would utilize all available traditional methods such that no special models are envisaged. Similarly, transfer is ineffectual unless the recipient has the capacity to put processes into operation. This in turn suggests a capacity building role for governments, a long-standing role. Finally, 'Devising measures to induce private sector owners of technology to transfer this technology to foreign organizations to the extent envisaged in Article 16 of the Biodiversity Convention will require considerable co-operation among all parties concerned'.

Australia[2]

The Australian Government supports technology transfer through a range of programmes, the most important of which is Australia's overseas aid programme. Assistance includes:

- strengthening environmental institutions within the Asia-Pacific region;
- development of a database of Australia's environment management capabilities and technologies; and
- linkage of market oriented enterprises in Australia and equivalent enterprises in selected developing countries.

These specific activities emphasize the Australian perspective that 'capacity building, including human resource development, technical skills and

---

[2] Sources for country reports include UNEP/CBD/SBSTTA/2/Inf.2 (1996), OECD (1996), and Lesser and Krattiger (1995).

services, institutional structures, management skills and protection of intellectual property [...] are all necessary to maximize and sustain the benefits of the technology transfer'. Note is made of the fact that the increasing commercialization of Australian public research institutions, including universities, makes them more client-focused and direct partners in technology transfer.

Germany

'Although technology development and use is mainly in the hands of the private sector, technology cooperation will be facilitated and enhanced by governments, e.g. through the establishment of the supportive political and economic framework.' Aspects would include:

- development aid programmes;
- demand-driven cooperation on a case-by-case basis; and
- use of the Clearing-house Mechanism to facilitate and promote co-operation.

Switzerland

The Government identified six possible roles, as follows:

- identify and list available technologies;
- provide information on access requirements;
- assist in private sector transfer;
- assist in the financing of technologies on concessionary terms;
- capacity building; and
- promote cooperation between developed/developing country private sector institutions, particularly on a pilot project basis.

The term 'concessionary funding' refers to payment in non-convertible ('soft') currencies. As many developing countries have non-convertible currencies, the availability of hard currency reserves becomes a limiting factor in purchases abroad, especially when the intent is for domestic use. A government can assist by accepting national currencies in exchange, and paying the supplying company in Swiss francs or other hard currency. In a related manner, governments can accept some of the private sector risks, such as possible liability. This is particularly a matter with corporate technology donations to developing countries, which, for Switzerland, include a rice *Bt* gene given by Ciba to IRRI and a marker gene useful for cassava, donated to CIAT by Sandoz.

In all cases, a distinction is noted with large countries and small firms. In general, firms will be willing to take more risks in transferring technologies to large countries due to the greater market potential. Conversely, small firms are less likely to undertake the complexities and risks of transferring technologies to developing countries. By extension, government assistance would be needed particularly in transferring technologies from small companies, and into smaller markets.

United Kingdom

Accomplishment of technology transfer is seen as involving:

- identification of relevant technologies;
- their development;
- access to them;
- their transfer; and
- successful assimilation and further development.

Focusing on the transfer aspects, small-scale projects are seen as the most appropriate, and then on a case-by-case basis. 'There is no single model which will suit all circumstances.' Practical access depends on financing, in which there is a public sector role. The public sector can also assist in the development of supporting capacity involving incentives for private involvement and human capacity for technology adaptation.

Developing countries

While the emphasis here is on the roles and plans of developed countries, the perspective of developing countries on their technology transfer needs for, and possible supply of, technology is also pertinent. Eight countries provided materials to the Secretariat, of which three are particularly useful here. *Bukina Faso*, noting the difficulties of poor countries in accessing technologies, particularly those controlled by IPR, identifies the financial and human resources needs of many countries in accessing technologies. Specific action is called from the CBD to assist in establishing appropriate systems (including protocols) of transfer and access.

From *Pakistan's* perspective, 'the technology issue was converted into a financial issue'. Specific financial needs are in research, education, and foreign training. The 'Green Revolution', despite the acknowledged problems, is identified as an effective technology transfer model. *Thailand* refers to the needs to promote and support scholarship, exchange experts, develop information networks, and establish institutions for supporting interactions between and among governments and the private sector.

Conclusions and Approaches

Technology is knowledge in a problem solving form, which implies adaptation for local use. Theory indicated that technologies will be traded internationally relatively heavily compared to other goods and services. The form of technology transfer depends on a number of firm and public sector decisions as well as on the attributes of the particular technologies (UNEP/CBD/SBSTTA/2/6, 1996, Par. 16). Key among these are production size economies and risk considerations.

At present, and in general, developed countries are net exporters, and developing countries net importers, of technologies. An increasing portion of those technologies affect environmental conservation directly or indirectly, or utilize genetic resources. Hence, the timely and efficient transfer of technology *is* essential to the achievement of the objectives of the Convention.

A review of the literature on technology transfer indicates significant potential. Further investigation, however, suggests significant problems as well, otherwise how can claims be made that there is much public sector technology of value to developing countries which has as yet not been adopted? The review indeed indicates that there are many factors which can impede the technology transfer process. Many/most of those factors are in the realm of individual economic decisions between sellers and buyers. Those decisions are only indirectly influenceable by seller country policy. Of secondary importance to the free flow of technologies is the policies of the recipient countries. Policies on regulations, ownership, taxes and IPR can have major influences on transfer, many of them negative. Policies of the supplying country would be of tertiary importance in influencing technology transfer, according to a distillation of the literature. Hence, it is not surprising the planned activities of developed countries as facilitators of technology transfer, the subject, of this chapter, are little developed. What though might those roles effectively be?

Virtually all knowledgeable observers, from inside and outside the CBD, governments, companies and the NGO community, recognize the need for *capacity building* and *funding*. These are traditional public sector roles which are no less appropriate in technology transfer. Funding relates primarily to making expensive technologies available. Capacity building, for its part, is seen as requiring a range of functions from general technical/scientific training to local adaptation to negotiation skills. Clearly, all are needed at some levels, and developing country governments (along with traditional bi- and multilateral donors) are the major available sources. Yet that community is experiencing 'donor fatigue' and can be expected to resist major new and additional appropriations no matter how justified. Thus some fresh approaches seem needed for establishing an effective programme in aid of technology transfer. It will be some time, if ever, before developing countries will be able to supply all their technology needs. Of course, multilateral groups such as the World Bank International Monetary Fund remain ready to assist countries in the implementation of optimal policies for attracting investment, but many developing countries, after often bitter past experiences, are resisting that aid (see Rapley, 1996).

At one level the onus will be on the recipient countries to establish joint programmes with the private sector. The SBSTTA refers to these as 'strategic alliances or partnerships' (UNEP/CBD/SBSTTA/2/6, 1996, Par. 53). The general concept is to establish an ongoing agreement which will provide, in addition to existing technologies access, longer term skills enhancement with the ultimate objective of being able to utilize effectively public sector technologies. For the present, many national public sectors lack the infrastructure and management for effective use so that technologies packaged with management must be purchased. The case of Piromal Enterprises, an Indian pharmaceutical holding company, which established an elaborate distribution and research partnership with Hoffman-LaRoche and the start-up Cytran, among others, is one example of what is possible (*Economist*, 1996b).

A requisite for successful partnerships is a better understanding and working relationship with the private sector. A certain wariness on the part of governments in dealing with sophisticated firms with sales exceeding many countries' GDPs is understandable, but does not resolve the matter that sustainable use and benefits sharing under the CBD will not succeed without a meaningful private sector role. Explanations are plentiful. The private sector only recently seems to be becoming aware of the significance of the CBD and of participating on an interactive basis. Also, some in the NGO community tended to vilify the private sector; a regrettable, documented example not directly related to the Convention is the admitted exaggeration of the evidence on the comparative risks of deep sea disposal of the oil storage tank *Brent Spar* by Greenpeace. There are adequate real issues to be presented without resorting to untruths.

The COPs can be more effective forums in bringing the two groups, government and private sector representatives, together. In the early years of the CBD there was limited private sector representation. That is changing slowly. The Executive Secretariat has been supportive of reaching out to the private sector and, while in the early stages, consideration is being given to the private sector role (e.g. SSRC, 1996, Annex II). While these effects are in the early stages, a more formal role for the private sector still being in the planning stages, the attempt is underway and should be applauded. One key role then for developed country governments is to encourage and assist the involvement of the private sector in the CBD (UNEP/CBD/SBSTTA/2/6, 1996, Par. 76). The two parties need not like or trust each other fully, but they must learn selective cooperation.

There is also a significant and on-going need for specific technology transfer assistance like that provided by the International Service for the Acquisition of Agri-biotech Applications (ISAAA), the Agricultural Biotechnology Support Program (ABSP), and the Consultative Group for International Agricultural Research (CGIAR). Such groups can identify and direct funding to small scale projects. They can arrange targeted training. But perhaps most significantly they are a buffer between the unknowns of the private sector and developing country representatives. Indeed, the roles of these groups can expand to provide information on standard terms and conditions as well as recent agreements (disguised as appropriate as to source). Specific roles can be debated; the general point is that commercially derived technology transfer will develop slowly in the complex and wary atmosphere under the CBD. Well-targeted project-level assistance can advance the process, as an addition to purely commercial transactions (see, for example, Krattiger, 1997).

That said, it can be argued that ISAAA and ABSP have made limited progress on the transfer of individual technologies. This is likely due to the reality of working with programmes and agencies, not the final users. Moreover, the number of projects possible with these small institutions is limited so that even doubling the rule-of-thumb success rate to one in five technologies with widespread adoption means the direct impacts will be minimal. Indirect support through capacity building on biosafety, IPR, and negotiations is probably a more effective function.

Beyond these general public sector roles of capacity building and facilitation some more specific critical functions can be identified as follows:

Risk shifting

The transfer of technologies is by its nature a risky enterprise, and the more so if involving living organisms as with some forms of biotechnology. Risks are of several dimensions, including financial risks of loss, liability risks, and risks of a public relations nature. All transactions involve risk, so the particular point being made here is that the nature and extent of the risks are unknown. This is a deterrent to trading, especially if the market potential is small. In cases where corporations are considering donations for use by small farmers, these risks could end the process. A possible action is therefore directed to shifting risks away from the firm to other entities, possibly including the government, as follows:

Financial

Financial risks are relatively easy to compensate for as they are clearly defined. There are three basic approaches: (i) to subsidize transactions thereby reducing private costs and the potential and scope of loss; (ii) to make up losses after the fact; or (iii) to spread risk through joint ownership. Subsidies can be approached through loan guarantees, guarantees which either reduce the cost of capital by providing a secure creditor or which base repayment on conditions such as the success of the project. Loan performance requirements are really a non-symmetric joint venture in which the government would not benefit from success but would share the costs in the event of a loss.

Liability

There are two basic approaches to reducing liability: (i) providing insurance in case of mishaps and (ii) transferring ownership, as liability typically goes with ownership. Presently there are major insurance firms providing various forms of environmental insurance; the other option for large firms is self-insurance. These are highly technical matters which are not conducive to government involvement. However, it should be noted that insurance operates on a pooling (averaging) basis. While a risk pool covering applications not provided by private insurers for a single country would provide too small a pool in most instances, it is possible to consider a multi-country pool.

The matter of shifting ownership is somewhat more straightforward. The most appropriate owner in the case of a donated technology would be the recipient government or agency thereof. With ownership at the time of any accident clearly that of the government, it would be difficult to ascribe responsibility to a previous owner. Alternatively, an agency of the government could become the owner of record, in either case the government making the necessary agreements.

Concessional terms

The term 'concessional' when referring to trade matters as in Article 16(2) is typically taken to mean payment in national, non-convertible currencies. The concept is that hard currencies are in shorter supply so that easing that constraint will make items heretofore unaffordable available. Here, the suggestion is to sell donated technologies in local currencies and use those funds in cooperation with an NGO for national biodiversity conservation. That process would have the advantage of forming a direct connection between biotechnology transfer and genetic resource conservation, something implied by the side-by-side positioning of Articles 15 and 16 but not yet achieved in practice.

Corporations have on occasion been willing to donate technologies to developing countries. Alternatively, public research institutions make technologies available without charge, as Cornell University has granted, via ISAAA, rights for small farmer use of virus resistance technologies for papaya to Brazil. As a final option, technologies could be purchased with a governmental or other pool of funds and donated to a country for sales in national currencies.

Reducing management requirements

The number of agbiotech applications reaching the final approval stage combined with reductions in numbers of midlevel executives as a result of corporate downsizing have placed great pressure on remaining managers.

One of their most scarce commodities is time, yet biotechnology transfer, being still a novel practice, is inordinately time consuming. A technology for controlling papaya ring spot virus, for example, is actually five separate technologies, of which the resistance gene is owned by Cornell, two vectors by Asgrow (which was recently sold), and the 'gene gun' was used, itself licensed by Cornell to DuPont to Agricetis which added its own improvement patents. Transferring such a complex bundle of technologies involves a major negotiating task, all the more because none of the standard practices of agreements have been developed. For example, how are royalties to be adjusted if the effectiveness of the resistance declines over time? Finally, Cornell donated its technology for small farmer use but not for commercial/export production. How in the field are the two sectors to be separated?

As a practical matter, donations and arrangements for small markets simply rarely get completed. In many regards, this is the major impediment to biotechnology transfer for small farmers/markets. Alleviating the situation, expanding the movement of private sector technology, will require finding substitutes for corporate management time. A major contribution would be the standardization of agreements, the creation of a model contract. This may not be used in its entirety but can function as the basis for negotiations with the longer term objective of routinizing exchange arrangements. Developing these agreements will be an educational as well as a legal/commercial process. Education will be particularly involved in helping the developing country partners understand the contract negotiating process, and consequences of contract terms. A significant amount of private firm management time now is indirectly spent in interpreting for the developing country negotiators the meaning of the terminology. Yet clearly that dual role is problematic for both parties, indicating the need for a third party.

Chapter 9

Roles of Intellectual Property Rights in Achieving Technology Transfer Objectives Under Article 16

Intellectual property rights (IPR) are identified in three of the five clauses of Article 16. The matter of IPR and the Convention, however, gained international attention when the US Bush administration did not sign the 1992 Convention in Rio, citing inadequate protection of IPR (Pistorius, 1992). The subsequent Clinton administration interpreted the matter differently – an interpretation supported by a major biotechnology corporation (Duesing, 1992) – and signed in 1993. That incident, as well as the general controversy surrounding IPR in the developed/developing country context, none the less continues to focus much international attention on the IPR references in the Convention.

The intent of this chapter is the presentation of information on the roles and record of IPR as a component of technology transfer. National roles in technology transfer are discussed in Chapter 8, and IPR as applied to claiming ownership in general in Chapter 2, and ownership of traditional/indigenous knowledge in particular in Chapter 7. Some overlap clearly exists among these discussions as IPR serves multiple roles, but to the degree possible the focus here is solely on IPR and technology transfer[1].

---

[1] For example, such suggestions as the mandatory reporting of origins of genetic materials on plant applications is discussed in Chapter 7 and the use of MTA for any special access considerations for the country of origin in Chapter 2.

IPR, considered in terms of direct references, occupies only a small part of the CBD. Three of the five clauses of Article 16 refer to IPR, Article 16(2) mandating that access and transfer of technologies must be consistent with 'adequate and effective' protection of IPR. Article 16(3) includes within the same sentence 'mutually agreed terms' and 'intellectual property rights', a further indication that the rights bestowed by IPR must be respected. Finally, Parties are enjoined from allowing IPR to impede the Convention, 'ensure that such [IPR] rights are supportive of and do not run counter to its [CBD's] objectives' (Article 16(4)). Indirectly, connections can be made with Articles 8 (j) (indigenous knowledge), 11 (creation of incentives), 12(c) (research cooperation), 15 (access to genetic resources), 18(1) (international technical and scientific cooperation), 19(1) (effective participation in biotechnology research), and 20(1) (provide financial support and incentives).

More indicative of IPR under the Convention is the amount of COP time devoted to the subject. Decision II/12 made three specific requests of the Executive Secretariat:

1. Invite the Secretariat of the World Trade Organization to prepare a paper for the COP which 'identifies the synergies and relationship between the objectives of the CBD and the TRIP agreement ...';
2. Consult with stakeholders to gain an understanding of the 'needs and concerns ...';
3. Undertake a preliminary study analysing the 'impact of IPR systems on the conservation and sustainable use of biological diversity and the equitable sharing of benefits ...'.

The final request led to the preparation of 'The Impacts of Intellectual Property Rights Systems on the Conservation and Sustainable Use of Biological Diversity and on the Equitable Sharing of Benefits from Its Use' (UNEP/CBD/COP/3/23, 1996). Request (1) provided for the preparation of 'The Convention on Biological Diversity and the Agreement on Trade-Related Intellectual Property Rights (TRIPs): Relationships and Synergies' (UNEP/CBD/COP/3/22, 1996). That topic is at a very preliminary stage of development and will be pursued further at COP4. While the Parties are likely to have little sway over the requirements of the TRIPs agreement, 'According to some delegates [...] it is appropriate for the CBD to have an opinion on environmental and socio-cultural aspects of IPR and to communicate this opinion to relevant fora' (*Earth Negotiations Bulletin*, 1996).

Yet this too masks the controversy surrounding IPR matters right back to the negotiation of the CBD. Those issues are reviewed on pp. 170–173, following a general consideration of IPR and technology transfer (pp. 162–164) and forms and operation of traditional IPR systems (pp. 164–167). The current status of IPR worldwide is discussed on pp. 174–177, and the chapter concludes with an overall assessment.

Conceptualization of the Role of IPR in Technology Transfer

There are two fundamental justifications for IPR systems, known as the personal property or 'natural law' and economic incentive approaches. The personal property approach is based on Locke's concept of a right to property being conferred by God upon all men in common (see Hughes, 1988; Thompson, 1992). This is in contradistinction to the absolute power of sovereigns. That concept of property though applies to common property, but what of personal property? Locke handles that matter by introducing the idea of labour, 'he that mixed his labour with and joined it to something that is his own, and thereby makes it his property'. Underlying is a view that a free person controls his labour, and a loss of the right to the product of that labour implies a loss of freedom. Property rights, including IPR, are thus a means of protecting freedom.

The economic incentive approach is more pragmatic, and less philosophical (the classical explanation is Machlup, 1958). It recognizes that the inventor assumes time and other costs associated with the creation process such that she/he could never compete on equal terms with copiers whose costs, minus the creation process, are lower. Hence the creator will always be undersold and has no incentive to invest. IPR legislation redresses the balance, at least in part, by prohibiting direct copying so long as the protection is in effect.

To be more specific, the invention process has been divided into three components: discovery, development, and commercialization. The discovery process itself seems to function more by creative drive, or mere luck, and hence is somewhat removed from financial incentives. Development and commercialization, however, are the lengthy and costly processes of turning an idea, an insight, into a marketable product. Work at these stages is very responsive to incentives and can be considered as the real target of IPR systems (Jewkes *et al.*, 1969; Chaps 15 and 16).

Of these competing concepts, which is the operable one for current systems? An insight can be gained from the authorizing legislation in the United States. There in the Constitution (Article 1, Sec. 8, emphasis added), it states, 'The Congress shall have the power ... *To promote the progress of science and the useful arts*, by securing for limited times to authors and inventors the exclusive right to their respective writings and discoveries'. This terminology has quite conclusively been identified as fostering economic incentives (Anderfelt, 1971).

The key function of IPR is therefore providing *incentives* for investment in the creative process, and in particular the transformation of basic insights into marketable products. These incentives are most applicable to private entities but have been used increasingly by the public sector as a source for generating research funding.

When considering the incentive effects, it is important to recognize what privileges IPR do and do not provide. They do not assure a return; indeed only up to 15% (and often far less) of patents are ever commercialized (Nogues, 1989). They do not necessarily permit the use/practice of the creation. That is often controlled by regulation (biosafety) or even other patents. All that IPR allow is the right to exclude others from use – what can be called negative rights. All financial rewards come from market sales. Hence key factors such as the breadth (scope) of protection and enforcement are critical in determining the practical value of IPR.

The Convention Article 16, however, is written as if pertinent technologies already exist, and that the relevant matter is access to those technologies. To a degree, that implicit position begs the matter because technologies must typically be adapted for localized needs, a process in itself often requiring IPR protection. However, it does draw attention to a second and generally less recognized aspect of IPR – its ramifications for *access* to protected creations (see Primo-Braga, 1989; Lesser, 1991, Chap. 4; Gutterman, 1993). It is in this context that the Convention's focus on technology transfer can be understood.

In general, it is difficult to predict the effect of IPR on access. On the one hand, private firms would understandably be unwilling to transfer technologies to countries where IPR legislation was considered inadequate, and all the more unwilling if a possibility of fostering competition in a third market exists. In that context, it needs to be noted that IPRs are strictly national law; it applies only where available and secured, otherwise not. Hence, while IPR owners may sometimes refer to the use of their technologies in countries where patents are not held as 'piracy', strictly speaking that use would be perfectly legal.

To continue, companies are reluctant to transfer technologies to countries with inadequate IPR (OECD, 1996). That applies particularly to easily copied technologies, such as pharmaceuticals and open-pollinated plants. On the other hand, it is difficult to prevent access to those products, seeds especially. Hence, some companies may seek to license those technologies none the less using the strategy that some return is preferable to none. The balance between those considerations, that is to say the net effect on transfer, is not well documented in the literature. Overall, it seems likely that firms will delay transfer to IPR-deficit countries, if only by making them a lower priority than areas offering stronger protection.

A different set of considerations arises when technologies are difficult to copy. This would apply, for example, to many complex processes. In those cases, copying is often slow and inefficient, the latter because the developers do not publicly reveal many of the technical aspects. The technicalities rather would be provided under a licensing agreement. That use of secrecy indicates IPR are but one means of protection; secrecy can often be used as well. Firms, for example, may transfer only to subsidiaries, keeping the technology within the firm, or may work only under contract (see Primo-Braga, 1995). Plantations would be a means of maintaining control over agricultural technologies. The net effect of weak or absent IPR can then, in general, be anticipated to be *delayed* access to new technologies, or *access under limited conditions*, such as only through subsidiaries or contractees. Thus, IPR is often an aid to technology transfer, especially for easily copied items, but in itself is neither necessary nor sufficient for technology transfer. In individual cases numerous considerations come to bear, so predicting the outcome would be very difficult. These matters are treated in more detail on pp. 169–173.

Forms and Operation of Traditional IPR Systems

The purpose of this section is a more detailed examination of IPR forms and functions as they apply to technology transfer under the Convention (see also Chapter 2).

Intellectual property rights (IPR) systems traditionally include five forms of legislation: patents, Plant Breeders' Rights (PBR), copyright, trademarks, and trade secrets. Of those, patents, PBR, trademarks, and trade secrets, both singly and combined, are generally directly applicable to applications under the Convention and receive attention here. Other more recent and specific forms of IPR exist as well, including those for maskworks (computer chips), which are described in Chapter 7 as regards indigenous knowledge.

Patents

Patents, like other forms of IPR, operate as a balance between the inventor and society. Society grants a temporary, partial monopoly to the inventor. Temporary refers to the duration of protection, generally 20 years; and partial describes the scope of protection, the degree of difference required before a related development is not covered by the patent. What society receives in exchange is more investment than it is expected would otherwise occur and the revealing (disclosure) of the invention. A typical patentability requirement is disclosure 'in such full, clear and concise and exact terms as to enable any person skilled in the art or science to which it appertains … to make, construct, compound or use it'. When for living matter a written description is judged insufficient, a deposit may be required (Straus and Moufang, 1990). Disclosure not only permits competition soon after a patent lapses but also provides a storehouse of technical knowledge which would not otherwise exist.

An additional patentability requirement is novelty – the invention must not be previously known. Finally, the invention must not be an obvious extension of what already exists; this is known as the non-obviousness or inventive step requirement. Hence it is not possible to patent just anything; the requirements are specific and exacting. Moreover, there must be human intervention in the inventive process. The mere identification of something existing in nature (technically known as discovery as opposed to patentable inventions) would not be sufficient for a patent. Examples of human intervention are the purification of a strain of microbes, or the identification of an especially rare rose mutant.

It should be further noted that, to identify a specific hypothetical case, a patent would not apply to all rice. Rather, the application would apply to rice with certain characteristics, such as the built-in insecticide *Bacillus thuringiensis*. Product-by-process potato, however, could conceivably be inclusive of all varieties produced using a patented breeding process.

Plant Breeders' Rights

Plant Breeders' Rights is a specialized patent-like system for cultivated plants. PBR were first systematized in 1961 under the International Union for the Protection of New Varieties of Plants (UPOV). At the time of writing, there are 32 members, including only six developing countries (South Africa, Argentina, Chile, Colombia, Paraguay and Uruguay). Others such as India, Pakistan, Mexico, and the Philippines could join in the near future. Still others, minimally Kenya, Taiwan, and Peru, have national laws, but the degree of their implementation is not known. UPOV membership among other steps requires that signatories adopt national legislation along the lines of that Convention.

In place of the novelty, non-obviousness, and utility requirements of patent law, PBR uses distinctness, uniformity, and stability (DUS). Uniformity and stability are measures of reproducibility true-to-form, respectively among specimens within a planting and intergenerationally. The principal test then is distinctness, that the variety be 'clearly distinguishable from all' known varieties. The DUS attributes are (except in the USA) generally measured in growouts of the planting materials.

PBR are further distinguishable from patents by the allowance of so-called 'farmers' privilege' and 'research exemption', sometimes called 'breeders' privilege'. The farmers' privilege is the right to hold materials as a seed source for subsequent seasons (farmer-saved-seed or bin competition), something which would generally be an infringement with patented materials. The research exemption refers to the right to use protected materials as the basis for developing a new variety or other research use. Research or experimentation under patents is not as well defined but is generally believed to be fairly broad.

Because of these differences, PBR are generally considered to provide less protection than patents. They also apply to the whole plant or the propagating materials thereof. What they do not protect is the unique characteristic (the distinguishing characteristic) of the variety. For that reason, no real protection is provided for a variety with a bioengineered gene which legally can be removed and used in another variety or with another distinguishing attribute added.

That situation will change under the 1991 UPOV text which in Article 14(5) allows for dependency. While experimental use remains unrestricted, a variety determined to be dependent on an 'initial variety' itself protected by PBR can not be commercialized without the permission of the owner of the initial variety. To be dependent, a variety must be 'predominately derived'. It may be obtained by selection, backcrossing, genetic transformation, or other specifically identified procedures and contribute a minority share of the variety's genetic materials. The actual interpretation of these general concepts is and likely will remain unclear until there have been some actual cases (see Rasmussen, 1990). The 1991 text further allows (but does not require) countries to restrict the farmers' privilege. To date, the USA will not do so, but the European Union considered limits on larger farm business.

Trade secrets

Trade secrets, to describe them in their simplest terms, assist in the maintenance of secrets by imposing penalties (the recovering of costs) when information held as secret is improperly acquired or used. Examples of trade secrets include customer lists and practices for improving the efficiency of a breeding process. An employee going to work for a competitor typically would be enjoined from revealing sensitive information for a specified period. Unlike patents and the like, no formal application procedure is needed for a trade secret; rather the information must have some commercial value, and an effort made to keep it secret. As long as these conditions are met, protection can be permanent. (For a description of the law and its application in the USA, see Coe (1994).)

Within agriculture, F-1 hybrids may be considered a form of trade secrets. As long as the crosses and/or the pure lines are protected, the product is difficult to copy. However, the self-reproducible nature of most living organisms precludes a major role for agricultural products. In other technological areas, trade secrets may substitute for, or complement, patents and PBR. When a product or process is difficult to copy, then trade secrets can be a substitute.

Trademarks

Trademarks are the reservation of a word, symbol, or phrase in association with a product or service. In effect the trademark name represents the product to consumers, justifying an investment in its identification. From a theoretical and economic perspective, trademarks assist customers in identifying products of consistent (and often high) quality. Trademarks are permanent as long as they remain in use, are identified as such, and do not acquire a generic connotation. Often a trademark, such as Coca-Cola, is the most valuable asset of a corporation. Within agriculture, trademarks can be associated with products at the firm level (Pioneer Hi-Bred), or individual products such as the FlavrSavr tomato. Note that the tomato variety is also patented, so the two forms of IPR are, in that instance, complementary. At the plant variety level, the role, however, could be more of a substitute than complement. Indeed, because of the farmers' privilege and research exemption under PBR, Lesser (1987b) has previously argued that in the USA, the PBR law really protects the variety name rather than the germplasm itself. Hence, there is a degree of substitutability between trademarks and PBR. The same would not apply to patents because of the emphasis there on identified novel characteristics rather than on the entire plant.

Evidence on the Implications of IPR

As was noted above, IPR are primarily a form of economic policy intended to advance the production and use of new products and technologies. That, however, is but the promise. This section explores the available information on what is known about the practical results of the legislation in the areas of investment and access. The information is anything but complete, but it is all that is presently available for planning purposes.

Investment (R&D)

Since a (the) major justification for IPR is the attraction of funds for research and development (R&D), it is a reasonable question to enquire about the evidence indicating actual experiences. For patents covering all technologies what is known is inconclusive. The analytical complication is largely methodological, attempting to determine what would have happened in the absence of the legislation. Additionally, for many technologies, other forms of protection can serve as at least a partial substitute for patents. Indeed, surveys of all business leaders typically place a low ranking on patents as a stimulant for R&D investment (Nogues, 1990, pp. 11–14). However, for biotechnology – many of the kinds of inventions envisioned for transfer under the CBD – industry representatives rank IPR as 'essential or of major importance' (OECD, 1996).

When specific sectors are examined, the results become more definitive. In general it is recognized that patent protection is especially important for pharmaceutical products and for living organisms. Both are relatively expensive to develop and easy to copy. A major cost is that of satisfying regulatory requirements. For pharmaceuticals in the USA, human trials are said to use the bulk of the $250 million per product development cost, and the preparation of a food safety dossier for a genetically engineered food costs around $1 million. One source of information on the role of patents is the implications of the *removal* of protection. In India, pharmaceutical R&D fell 40% from 1964–70 to 1980–81, something Deolalikar and Evenson (1990, p. 237) attribute to the weakening of patent protection in 1970.

An ancillary point, and one particularly relevant to agricultural applications, is that of adaptive research. Deolalikar and Evenson (1990, p. 251), again referring to the case of India, conclude, 'If anything, the relationship that is often observed is one of complementarity'. In Evenson's view (1988, p. 152), 'Indirect transfer does not take place without research capacity in the destination country'.

A number of more formal economic studies have been conducted on aspects of the patent system, such components as optimal duration and the consequence of the 'winner take all' approach (review in Primo-Braga, 1990). Overall, as might be expected, these issues are very sector specific, and general studies lead to inconclusive results with limited policy implications. However, indications are that patenting and R&D are not dominated by major firms. Rather, medium sized entrepreneurial firms which are dependent on technological advantages for their market position are the market leaders.

Overall, PBR are relatively much more recent and sector specific than patents, which eases the methodological problems in evaluating the impacts. The major study was conducted in the USA in 1980, a decade into the Plant Variety Protection Act (Butler and Marion, 1985). When considering the results, it should be recognized that the USA interpretation of not requiring objective standards for performance claims means the scope of protection there is relatively narrow (see Lesser, 1987c). Despite these caveats, it was found that PBR did have a significant impact on private investment and numbers of private breeders, especially for soybeans. Those results have been confirmed by other observers (e.g. Brim, 1987, Tables 3 and 5).

More recently for the USA, there are some suggestions that the initial investments by the private sector were over responses, that the actual profits are insufficient to maintain the current investments. The premium for certificates for soybeans in New York State was placed at only 2.3% (Lesser, 1994c), which is consistent with that position. Hansen and Knudson (1996), however, concluded, again for soybeans, that PBR holders are able to price to capture the varietal value of harvest saved as seed (indirect appropriation), suggesting adequate property right protection. Some companies including Stine Seed Company, none the less, have not been pursuing PBR in favour of sales agreements. A typical agreement would read in part, 'Purchaser hereby acknowledges that the production from the Stine Brand Seeds herein sold ... will not be used or sold for seed, breeding, or any variety improvement purposes'. The level of enforceability of these provisions in the USA, not to mention other countries, is not known.

Recently, limited information has begun to appear on the operation of PBR in other countries. A graphic plot of new variety registrations in South Africa indicates a notable increase following the adoption of PBR in 1976 (van der Walt, 1994, Table 1). Similarly, the Argentine private sector increased their investments in plant breeding, a provisional study shows, but only after the law was enforced (Jaffé and van Wijk, 1995). Hence all available information is consistent with the theoretical expectations that increased IPR does indeed lead to greater investment, especially for easily copied products and where capacity exists. The more relevant, and difficult, question for the Biodiversity Convention is the implications for access.

Access

The conclusion that PBR leads to greater internal investment in breeding expenditures leaves some ambiguity regarding its effects on access. Access conceivably could be enhanced, supplanted by recipient country investments or could remain unaffected. Many of these are long-term issues for which a few countries are just approaching the initial stages. None the less there are some bits of information which do suggest that the presence of PBR does indeed enhance access.

A strong motivation for the recent (1990) adoption of PBR by Canada was access to improved, protected potato varieties from Holland. As well, within Canada there was a reluctance to export varieties to the USA because of the concern that they would be transported back into Canada (Cooper, 1984, p. 47). Young (1989) concluded:

> Some private varieties should be available for use in Canada even though they may be created elsewhere as part of the larger plant breeding program, and a consequence of no breeders' rights legislation is a restriction on the availability of such private varieties.

Similarly, cut flower producers experiencing difficulties with accessing new varieties were major proponents of the Colombian national law and subsequent accession to UPOV. Uruguay adopted PBR largely to prevent trade disruptions with Argentina, to which its economy is closely tied (Jaffé and van Wijk, 1995, p. 20). Overall, Juma and Ojwang (1989, p. 153) recognized that the greatest restriction on the exchange of improved varieties is with countries outside the UPOV system.

The evidence for access under patents is more diffuse because of the range of technologies affected. However, Pray (1986, p. 50) found that 'The likelihood of copying [...] represent[s] major barriers to the introduction of new products in India by foreign companies', while McLeland and O'Toole (1987, p. 247) see limited patent protection as one reason why technology imported into Latin America is often outdated. The recent OECD survey (1996, p. 44) found 'Comprehensive intellectual property protection' to be a 'fundamental prerequisite not only to technology transfer agreements but also to attract and maintain ancillary investment ...'.

The transfer of agrobiotech products is only now becoming an issue as they reach the commercialization stage so the evidence regarding the implications of IPR for access has generally not yet appeared in written form. However, de Castro (quoted in Lesser 1991, p. 54) identified the lack of appropriate IPR protection as the reason why Monsanto has hesitated in sharing *Bt* cotton technology with Brazil. Also, Monsanto is said to have rejected a request from Egypt for herbicide-resistance technology because its proprietary herbicide could not be registered by name. For Duffey (1989), former head patent counsel for Monsanto, the situation is simple, 'When faced with a country that provides *inadequate* IP protection, multinationals will either take their products and technologies elsewhere or simply make available products that contain older technologies they can afford to lose'.

A recent unpublished survey of Swiss multinationals reinforced Duffey's point with the caveat that firms may risk new technologies in large and potentially lucrative markets (Lesser *et al.*, 1995). However, those are just the countries which often have the strongest IPR protection (Evenson, 1990, p. 352). That leaves small countries in a potentially difficult access situation (Lall, 1981). This evidence generally substantiates a position that IPR protection supports technology transfer/access as called for under Article 16 of the Convention and hence is in line with Convention objectives. However, the evidence is far from conclusive (UNEP/CBD/COP/3/22, 1996, Par. 58).

Concerns Expressed Regarding IPR Systems

Much critically sharp material dating back at least to the 1970s has been written about IPR and developing countries. UNCTAD (1975) drew particular focus. More recently, IPR has been criticized, especially by segments of the NGO community, as racist, imperialist and unethical (e.g. Sing Nijar and Yoke Ling, 1994; Shiva and Moser, 1995). Frequent calls have been made for restrictions, exemptions, or limitations (see UNEP/CBD/COP/3/22, 1996, Par. 35). Distilled, concerns can be categorized in three areas. One relates generally to the activities of firms, especially the potential that large firms, due both to financial resources and economies of scale, will dominate R&D, and the 'winner takes all' aspect of patenting which can lead to duplicative and wasteful research. The second is more specific to biodiversity and relates to changes in access to genetic resources under IPR. Finally are concerns expressed that the investments made are not applicable to the needs/opportunities of developing countries and are rather directed to the rich, industrialized countries. Issues regarding community rights and equity matters as applied to genetic resource use are discussed in Chapters 4 and 7.

Considering the first matter, a number of formal economic studies have been conducted on these and other aspects of the patent system (reviews in Scherer, 1980, Chap. 16; Primo-Braga, 1990). Overall, these issues are very sector specific, and general studies have had limited policy implications. However, indications are that patenting and R&D are not dominated by major firms. Rather, medium sized entrepreneurial firms which are dependent on technological advantages for their market position are the leaders in technology development, as can be seen, for example, in the decline in the position of IBM in the computer sector.

The 'patent race' is of possible concern because investors will establish their own research programmes in the hopes of beating a competitor to a patent application, but the investment of the second finisher, the one who does not get the patent, is largely wasted. However, by encouraging multiple approaches by different inventors to the same problem, leading on average to quicker success, the race to patent is recognized to have benefits as well as costs by bringing products to market earlier. For PBR this concern is presented in terms of 'cosmetic breeding', the production of multiple varieties with limited agronomic differences (particularly an issue under the USA system) (Claffey, 1981; Vellvé, 1989). However, the indirect effect is greater competition so the net consequences to users of higher prices (a result of wasteful investment) or lower prices (attributable to greater competition) are unclear.

Access to genetic resources is particularly an issue for plant breeding, for those resources are the basic input. The concern expressed is that IPR increases the value of genetic resources which will cause limitations on access. In a survey of public breeders in the USA, Butler and Marion (1985, p. 70) found the amount of material sent from universities to private firms increased under PBR, but firm-to-university transfers remained the same. The absence of an increase was attributed to PBR which gave the private sector an incentive to limit competition by maintaining tight control over promising materials. Broader sector surveys indicate delays in the release of research information generally associated with the industry need to preserve confidentiality until (and in some cases following) applying for a patent (Blumenthal *et al.*, 1996). Evidence from five Latin American countries is more convoluted due to the combined effects of PBR and public sector budget reductions. These factors led researchers to seek part of the commercial value of materials shared with the private sector. Overall, what has changed most has been the means of access, from unrestricted gifts to the use of research agreements claiming some future payments. In some instances, firms have found that this change to a formalized system actually enhanced access (van Wijk, 1995). More fundamentally, IPR probably does delay and restrict the transfer of knowledge compared with the common property system. The more appropriate comparison though is with the total amount of knowledge publicly available, a comparison which includes a consideration for the IPR-induced knowledge generation.

The technology irrelevance issue can be most easily described in terms of pharmaceutical products. With something like three-quarters of health care in developing countries coming from traditional sources, the role of encouraging the development of costly research-based pharmaceutical products by allowing patent protection has frequently been questioned. Moreover, many of the products are developed for the rich, temperate countries leaving a major unfilled need for the development of tropical disease treatments (see *Economist*, 1996c). There can be no serious counter position but that IPR is directed principally to private companies which are responsive to market opportunities. Hence it is unrealistic to expect that the mere existence of strong IPR legislation will increase R&D investments in all needed areas. The same may apply to technologies for environmental protection and the sustainable use of genetic resources, both of direct relevance to the objectives of the Convention. IPR provides incentives which are in proportion to market opportunities; when the market is small in terms of profit potential IPR will have a limited effect. Hence, investment effects are spotty and market/sector specific.

The same general reasoning applies to availability of improved seed by small, resource-poor farmers. Often IPR is criticized for not providing for this group (e.g. UNEP/CBD/COP/3/22, 1996, Paras 52 and 56). This is undoubtedly true – again, IPR operates on market incentives and does not function where there is no market – but in itself does not invalidate IPR. No single system can serve all needs. A more pertinent question is whether IPR injures resource-poor farmers or merely bypasses them. In many cases there is indirect harm if improved farming practices (including seed) raises output and reduces prices. Under that scenario, farmers whose costs do not decline will suffer from lower prices. On the other hand, urban consumers have benefited from falling prices so that the single technological change has caused a significant readjustment in welfare among farmers and farmers/non-farmers. Determining the net effects is complex, yet, clearly, considering only one group is inadequate.

The issues mentioned in the two preceding paragraphs are but specific cases of a more general concern that IPR is not appropriate for countries with a low level of technological capacity. According to that line of reasoning, countries with a limited industrial sector cannot produce world-class products suitable for IPR protection so that protection serves only to raise the costs of imports. As support, the recent extension of patent rights by some industrialized countries – Canada offered pharmaceutical patents only within the past decade – is used to argue that developing countries are not being accorded the same IPR-free period as were the industrialized nations. Clearly, countries with a small industrial sector (a group including several southern African nations, among others) will not participate in patenting to any degree, although countries like India and Brazil undoubtedly will.

To these concerns, two comments can be addressed. First, IPR, as is argued above, also serves as an aid in accessing current technologies (including biotechnologies) and to foster the development of national adoptive capacity. Second, if a country is at a low level of development, including a small domestic market, then companies are unlikely to go to the expense of patenting these anyway.

Some impediments to access are implicit under IPR. They function by allowing owners to prohibit use (see above). This includes the ability to prevent local production or importation from other sources when, for example, domestic prices are very high due to a small supply (UNEP/CBD/COP/3/22, 1996, Para. 56). Hence the accomplishment of the Article 16.5 mandate that 'The Contracting Parties ... shall cooperate ... to insure that IPR are supportive of and do not run counter to the Convention's objectives' is an interpretative issue. However, under TRIPs (see pp. 175–176) it may allow granting compulsory licenses 'predominately for the supply of the domestic market', subject to 'adequate remuneration' (MTN/FAII-AIC, Article 31). This gives governments some leeway to bypass onerous requests by patent holders. Reichman (1993, p. 204) concludes, '[these articles] explicitly entitle developing countries to assimilate concerns about economic development into these exceptions'.

van Wijk (1995) describes a particular example where an Argentine licensee was denied access to European markets for strawberry plants through a combination of patents and contractual agreements. The contract agreement giving the European firm exclusive rights to the European market was presumably the basis of the restriction. PBR was the enforcement mechanism allowing the prohibition of imports unauthorized by the inventor. Certainly that agreement could raise prices in Europe, but that is a distinct matter from access, as Europeans had a source of supply through the local licensee. Moreover, if the agreement led to excessive prices in Europe, that matter could be addressed directly through a compulsory license, although such actions are rare. What it is, is a restriction on the activities of the Argentine firm, something not uncommon in commercial agreements of this kind, with or without IPR.

Current Status of IPR Protection Worldwide

As of 1988, 53 countries statutorily excluded plants and 54 excluded animals from patent protection (WIPO, 1990, Annex II). These include the members of the European Patent Convention (EPC). The bulk of the other countries are developing nations, many of which have language similar to the EPC. Article 53(b) of the EPC excludes patents for 'plant or animal varieties and the essentially biological processes for the production of plants and animals'. Written in the pre-biotechnology days, the interpretation of that phrase has proven complex over the years. Several patents have been granted based on an interpretation that 'variety' refers to a variety in a 'fixed form' so a development which was applicable across multiple varieties could be patented. Most recently, the European Patent Office (EPO) appeals ruling on a Plant Genetics System patent rejected coverage for the plant and seeds suggesting plants would not be patentable (*Official Journal EPO*, 1995). However, the ruling is under appeal so that the significance of that decision will not become clear for some time.

As noted there are presently 32 members of UPOV, all but Argentina, Chile, Uruguay, Colombia, Paraguay and South Africa being developed countries. A number of additional countries have national PBR laws, including Taiwan and Kenya, among others. Details on the operation of those laws is limited. A number of countries are presently drafting legislation and/or have a likely goal of acceding to UPOV in the near future. This includes Mexico and Brazil as well as Pakistan, the Philippines, and India. Finally, another group of countries – including Indonesia and Malaysia – has begun the drafting process but seems likely to file in the more distant future.

Membership in a national convention, of course, standardizes the conditions of protection to a large degree. Standardization of patent and trademark laws is assured in part by the Paris Convention of 1883 with its 100-plus members. Among the key provisions are *national treatment* which requires that foreigners must be granted the same rights as nationals. Additionally, the *right of priority* stipulates that an application filed in any member country establishes that filing date for all other countries for a period of a year. The filing date is critical for the bulk of countries which follow the *first-to-file* system. The major difference is the United States which uses the *first-to-invent* procedure. The Paris Convention is administered by the World Intellectual Property Organization (WIPO), a specialized agency of the United Nations, which oversees administrative and harmonization responsibilities. The Paris Convention also sets limits on conditions for compulsory licenses, rules which allow third parties licences for patented technology.

The final major difference in worldwide patent laws is the form and duration of the *grace period*, the time between an invention being publicized and the initial filing of a patent application. These range from none in the

EPC member countries to 1 year in the USA (Lesser, 1987c). Recent efforts for further patent harmonization broke down, but some standardization is imposed by GATT, such as setting the patent duration at 20 years from the first filing.

For the immediate future, signatories to the Uruguay GATT Round, a list including essentially all countries of any size (except China and a few other communist economies), have made commitments to enhance IPR protection under the Trade-Related Aspects of Intellectual Property Rights component (see also UNEP/CBD/COP/3/23, 1996). The TRIPs agreement requires signature states, including some 70 developing countries, to provide for the following protection (MTN/FA II-A1C):

- Contracting parties shall provide for the protection of plant varieties by patents and/or by an effective *sui generis* system (Section 5, Article 27(3b)).
- Patents may be prohibited to protect *ordre public* or morality, provided there is a justification exceeding the mere prohibition in domestic law (Section 5, Article 27(2)).
- Plants and animals other than microorganisms and 'essentially biological processes for the production of plants and animals' may be excluded from protection (Section 5, Article 27(3b)).
- Compulsory licences may be issued in limited cases of due diligence to make a licensing agreement, adequate remuneration, and subject to judicial review (Section 5, Articles 30 and 31).
- For process patents, the burden of proof of infringement may in some specified circumstances be shifted to the defendant who must prove that the patented process was not used (Section 5, Article 34).
- Persons shall have the option of preventing others from using information of commercial value without permission so long as reasonable efforts have been made to keep that information secret (Section 7, Article 39).

Even with this legislation, restrictions will remain, for example, the 5 years (and up to 10 years depending on product and level of development of country, with further delays possible on approval) allowed for developing countries to adopt and implement the changes (Part VI, Articles 65 and 66). Moreover, similar terminology to 'plants and animals and essentially biological processes for the production of plants or animals' exists in the European Patent Convention (EPC Article 53(b)), but there reference is to 'plant or animal *varieties*' (emphasis added). The trend had been to interpret the exclusion narrowly so as to allow patenting which applies say to multiple varieties (see Crespi, 1992); but a European Patent Office decision in 1995 overturned that approach by defining genetically engineered plant varieties which then fall under the exclusion (*Official Journal EPO*, 1995). That decision is under ongoing review/reconsideration. How it will be interpreted with the new terminology is not known at this time, but in all likelihood patents for most life forms (except microorganisms) will be prohibited in at least some countries, but not biotechnology processes even when applied to living organisms. Additionally, note that countries may exclude patents which are contrary to '*ordre public*' or 'morality'. This terminology exactly parallels the EPC (Article 53(a)), and the European Patent Office has rejected an animal patent on those grounds. This increases the likelihood that some developing countries will exclude classes of inventions, living organisms in particular, based on moral objections. TRIPs (Section III) does require countries to adopt enforcement procedures which are 'fair and equitable', are 'reasoned' but 'not unnecessarily complicated or costly'.

All of this says that countries opposed to IPR may, under TRIPs, have to make limited changes, and none in the immediate future, unless and until countries are convinced change is in their benefit. And since the great bulk of countries presently with absent or constricted IPR legislation or enforcement are developing countries, the immediate implications of TRIPs for the Biodiversity Convention could be limited. However, within that general conclusion there are more specific matters to consider.

Plant Breeders' Rights

As regards PBR, TRIPs is quite specific – the allowance of either plant patents or a *sui generis* – system for plant varieties, or both. *Sui generis* means separate or independent, as in a distinct form of legal protection. This is widely interpreted to mean Plant Breeders' Rights as in one of the UPOV conventions. That is, UPOV membership, although no specific interpretation has to date been issued, would, in all likelihood, satisfy the commitment. The other option for countries is the adoption of a national PBR law, as presently exists in several countries.

Patents

Unlike PBR, countries have the right under TRIPs to exclude patents for plants (and animals). Considering the controversial nature of this matter, it seems many might do so, or more correctly, continue to do so. As of 1988, 54 national patent laws prohibited patenting plant varieties (WIPO, 1990; Annex II). Of course, a large step from the non-exclusion of plant patents to issuance remains. Even with the rapid evolution of views in India, it is difficult to imagine a plant patent emerging there any time soon.

PBR will not suffice to provide protection for genetically engineered plant traits for reasons which can be readily explained. Under the 1978 Act of UPOV text, any variety which is distinct in one (recognized) characteristic can receive protection. Thus, if a rice variety bioengineered for pest or disease resistance had improved yield added by a different firm, the improved variety, resistance and all, would be owned by the second firm. The dependence stipulation in the UPOV 1991 Convention text would allow more ownership control by the biotechnology firm (Article 14(5)). If the disease resistant variety were accorded 'initial variety' status, derivative varieties could not be commercialized without permission. However, nothing would prevent a firm from removing the responsible genes for transfer to another distinct variety. A combination of 1991 UPOV *and* patents on the genes themselves would seemingly provide protection similar to plant patents.

Conclusions and Approaches

Four forms of 'traditional' intellectual property rights legislation are applicable to protecting the kinds of technologies implied in Article 16 of the Convention, including biotechnologies. These four, which can be used separately or jointly, include patents, Plant Breeders' Rights, trademarks, and trade secrets. Each is intended for a particular function and as such has specific attributes and exacting granting requirements. Certainly it is not possible to protect all forms of innovation.

There are two conceptual justifications for intellectual property rights: the personal property argument and incentive mechanism use. Modern IPR systems typically emphasize the incentive factor. The available evidence generally supports that expectation; IPR do indeed encourage investment by the private sector, especially for easily-copied inventions. For less technologically advanced countries, or sectors in all countries, petty patents (utility models) have been shown to be useful for increasing investments in simpler mechanical inventions (see Chapter 6, p. 119). The evidence, however, is fragmentary and will not convince all readers, in part because IPR are but one means of protecting inventions. Maintaining physical control, or secrecy, can be used as well in many instances. According to the available evidence, IPR, nevertheless, is an important component of an incentive system.

The Convention, however, by emphasizing transfer, implicitly focuses on existing technologies so that it is the *access* role of IPR which would be more relevant and is the focus of this chapter. Access is less well studied, and difficult to project in general. Several competing forces exist depending on the specific technology, the fear of losing control of a technology vs. the impossibility of maintaining long term control, and the potential for using secrecy/physical control as a protective mechanism. Overall, limited IPR probably means delayed access with other, if less visible, costs. IPR systems have costs, royalty payments being the most obvious, but the costs of the absence of protection in terms of denied or delayed access must be determined on a case-by-case basis. Yet the available evidence is limited and the Parties of COP3 were well justified in encouraging the generation and communication of case studies. To reiterate the preceding point, IPR, according to available evidence, often enhances technology transfer, but it is neither a necessary nor sufficient condition for such exchanges.

Patent laws exist in over 100 countries; PBR in 32; no international tabulation of trade secret legislation exists. The raw figures, however, do not give a complete indication of the status of protection. Around 50 countries specifically exclude plants and animals from patent protection, although there has been some patenting made possible by a narrow interpretation of the exclusion. On the other hand, the number of countries with PBR legislation, especially members of UPOV, has been growing rapidly, with several (the Philippines, Mexico, India, Brazil among others) expected to join in the near future. This is due at least in part to the Trade-Related Aspects of Intellectual Property Rights (TRIPs) agreement under GATT which specifies certain minimal levels of protection be adopted within 5–10 years. In particular, signatory countries must have in place either PBR or patents (or both) for plants. Most at present are opting for PBR, although it seems to offer incomplete protection for bioengineered plants.

While a significant level of dissenting opinion does and will continue to exist, strengthened IPR seems the direction of the future. This conclusion coincides with national commitments under TRIPs, as well as national needs for access to technology. Sometimes it is pointed out, as a justification for delaying IPR legislation, that Switzerland did not allow patents until the era of World War I. Two responses seem appropriate: (i) many developing countries are now at a related economic stage when IPR is beneficial; and (ii) the world is a much different and more intertwined place than in the first quarter of the century. What was appropriate, even possible then, may not be so now.

Indeed, many countries, even long time sharp critics like Brazil and India, are strengthening IPR legislation, in a large measure for domestic reasons. The issue then arises, what forms of law and system are appropriate for the existing diversity of nations. Clearly, the level of detailed knowledge about the practical operation of these systems across a broad range of countries is limited. The question then seems not so much if, but how and by whom.

Since TRIPs, WTO has become a major world force regarding IPR, supplanting, by some interpretations, WIPO. As WIPO is dominated by developing country parties, it is sometimes seen as more democratic than is the WTO which can be influenced by the large trading countries. The Parties at COP3 requested the CBD Executive Secretary to contact WIPO regarding capacity building; WIPO staff (with UPOV) can likely assist in other areas as well.

The Parties also noted concerns regarding recent WIPO actions on IPR protection of databases. A modified treat was adopted by the WIPO Diplomatic Conference in December 1996 (WIPO, 1996) (see www.loc.gov/copyright/wipo6.html). Article 5, 'Compilations of Data (Databases)' reads as follows:

> Compilations of data or other material, in any form, which by reason of the selection or arrangement of their contents constitute intellectual creations, are protected as such. This protection does not extend to the data or the material itself and is without prejudice to any copyright subsisting in the data or material contained in the compilation.

The developing country concerns are that copyrighted databases will impede access to pertinent information in a manner related to that discussed above. The effects are difficult to assess at this early point. Countries can take some satisfaction that the underlying information (unless otherwise prohibited) will remain available, although additional cost and inconvenience will be required to access it (see Lawler, 1996).

A major issue for COP4 will be the implications of TRIPs on the objectives of the Convention. Given the fairly general nature of evidence regarding the effects of IPR, it is not immediately clear what additional substantive information will be available for discussion.

Within patents, a major choice exists between *examination* and *registration* systems. In the former the validity of an application is evaluated prior to granting, in the latter only when and if a suit is brought. Examination systems have many conceptual and practical benefits, but these can be easily eroded in fields like biotechnology where competent examiners are scarce. Registration is an expedient. For PBR, countries may choose to test (grow out) applicant varieties (as is done in the EU) or rely on competitors for assessment and policing (the USA approach). Each approach has benefits and costs to be considered by national policy makers. Dispute settlement is an additional critical matter. It can be complex for countries with limited numbers of trained judges and/or not-fully independent legal systems. An approach increasingly taken is dispute settlement through arbitration (Niblett, 1995). WIPO now operates an arbitration service, along with numerous other entities (WIPO, 1995; see list in Wetter and Priem, 1991). The latitude possible in compulsory licensing is broad, but the implications must be carefully considered. Trade secrets are in some regards the most complex to implement as they have cultural implications for the free exchange of information.

Overall, there is an apparent need for assisting countries in the consideration of alternatives for the extension/modification of IPR. Presently, WIPO provides training programmes for future examiners, but suggested here is a far broader and probably more controversial capacity building to include an understanding of the implications of IPR systems. Just which body or bodies would be responsible is not clear. Certainly, major multinational firms using IPR and major IPR offices, including the USA, Japan and EPO, should participate, but additional experiences and perspectives are needed as well.

Until IPR systems have been implemented on a broad scale, to some degree inadequate systems will frustrate the objectives of the CBD. Achieving an operable level of IPR across diverse national economies will require considerable capacity building. TRIPs has set the clock in motion so that there is urgency to be progressing with that capacity enhancement. IPR legislation, over the years, has been modified on average every 20 years. Now is the opportunity to get it right from the start; otherwise, the wait will be a long one.

Once operational, some countries/institutions will need assistance in financing access. This can take two forms. First is the cost of procuring IPR protection. In many cases those costs are subserved by a corporate partner in an MTA, as with Merck/INBio. It is not necessary to be the direct IPR holder to benefit. Otherwise, a rotating national fund can be established for securing IPR. Conceivably the Financial Mechanism of the Convention (Article 21) could endow national or international funds for the purpose. Second, the additional costs of access to IPR-protected inventions needed to achieve the Convention's objectives could also be supported by the Financial Mechanism – the Mechanism's operation and priorities are incompletely structured at the time of writing. In many cases, royalties will be a small part of larger projects/activities.

Chapter 10

Summary and Conclusions

The preface to this book identifies an underlying precept of this work: the significance of economic factors as they affect the use and conservation of biological diversity as set out in the Convention on Biological Diversity. Here, pragmatic economic theory and practice are applied to seven key Convention articles with the intention of advancing aspects of the use of biological diversity, including equity. The objective of conservation is addressed only indirectly through the expectation that 'what is of value is protected'. Other mechanisms for conservation incorporated in the Convention but lying outside the direct operation of markets are beyond the objectives here.

The approaches to equitable use identified here can be summarized under eight subject headings:

- material transfer agreements;
- access legislation;
- valuation and equity;
- prior informed consent;
- materials held *ex situ;*
- property rights for indigenous knowledge;
- technology transfer – the roles of governments; and
- technology transfer – the roles of intellectual property rights.

To a degree the divisions among these are arbitrary, as they all relate to different aspects of the same issues. Partitioning is useful none the less for enhancing comprehension and implementation.

Material Transfer Agreements (Article 15)

By many appearances, the 'common heritage' approach to the sharing of the world's genetic resources without compensation to, or even control by, the source countries is approaching an end. The underlying causal factor appears to be technology, including biotechnology, as it enhances the commercial value of genetic resources. Casual observation, buttressed by economic theory, suggests that property rights systems often develop when resources acquire commercial value, and so it is with genetic materials.

First came property rights for finished materials (outputs) in the form of seeds, medicines, and the like. These property rights advanced calls for means to claim compensation for genetic resources, the raw material for many new technologies and products. An assessment of intellectual property rights mechanisms by which the *products* of biotechnology are protected, however, indicates they are seldom applicable to the *inputs.* Suppliers understandably consider it unfair that their material is not accorded commercial value while the resultant products can be quite profitable. Hence, genetic resource suppliers are without a general mechanism other than secrecy for protecting their property and, thus, often resort to secrecy. In terms of the Convention, this means restricted access, counter to the spirit of the Convention. Given this situation, and the motivation by some owners of genetic resources to realize cash value, use agreements may be achieved through material transfer agreements (MTAs).

MTAs are quite complex, especially at this early stage when no standard form has emerged, and are thus costly to negotiate, monitor, and interpret. Successful use of MTAs would seem to require the following elements:

- identified objectives (payment, value added industry, etc.);
- skilled negotiators;
- information on prevailing terms;
- trust and a spirit of partnership; and (not to be overlooked)
- access legislation.

The first listed is the responsibility of the suppliers, whether government or local community, and requires both insights into expectations of compensation and a realistic understanding of potential and business realities. The subsequent element will take time to develop, through capacity building. There have been multiple calls for capacity building support for negotiating, but no coordinating entity has as yet been identified. The function of that entity (provisionally termed the 'Facilitator') could expand to provide public information on agreement terms. Trust and partnership – for these complex agreements can tie the parties together over sustained periods – will evolve, but agreements should not be undertaken with a complete absence of trust, as no agreement can be structured which permits verification of every aspect.

Properly structured MTAs, however, can assist in enhancing trust and in apportioning risk between genetic resources sellers and buyers. In general, a high initial payment signals trustworthiness of the buyer; buyers may then wish the payment to be made in equipment rather than cash to increase the likelihood that the seller will meet the terms of the agreement.

Proportionally greater royalty payments (compared with initial payments) increase the expected eventual payment to sellers, but at the cost of higher risk of no payments. Royalty payments also reduce the need for information on market prospects, helping to reduce a negotiating disadvantage of sellers. From another perspective, high initial payments may be seen as loans from the buyer, potentially valuable depending on the implicit level of interest, but probably reducing eventual total payments.

Finally there is access legislation which makes the completion of MTAs a requirement. The mere existence of the CBD with its 'sovereigning rights' principle is insufficient; countries must adopt national legislation controlling access.

Needs and approaches

- Identification of standardized forms of MTAs for access to genetic resources with benefit sharing.
- Capacity building in interpreting and negotiating MTAs.
- Sharing of information on terms and conditions.
- More involvement of the private sector with the CBD process.
- Funding for capacity building and oversight body with identification of a responsible body (the 'Facilitator').
- Continuation of efforts by the CBD Executive Secretary and by private sector trade organizations in particular to attract more private sector involvement in the CBD process.

Access legislation (Article 15)

Access legislation is fundamental to the use aspects of Article 15. It serves on the one hand to prohibit unauthorized use of biodiversity by requiring a permit for collection and related activities. On the other hand, it functions as the mechanism for implementing benefit sharing and prior informed consent requirements. As such, it must operate on multiple levels, nationally and locally. Thus access legislation is not simple, yet undue complexity should be avoided as encumbering the process and raising costs.

To date, only two entities – the Philippines and the Andean Pact – have adopted complete access legislation. (The Pact legislation requires enabling legislation in the member states of Bolivia, Peru, Ecuador, Colombia, and Venezuela, of which only Ecuador has acted.) Several other countries have drafts in place (Brazil, Fiji), while Australia has advanced the process but must contend with the additional complexities of a federal (divided national/state authority) government system. Given this, it is no surprise that there are among these no obvious models for adoption elsewhere, but they provide a good starting place for succeeding legislation.

Concise and functional access legislation would seem to require the following components:

1.   Scope: What is included/excluded from the legislation.
Recommended: Be inclusive while excluding specific materials; leave the extent of derivatives included to be decided in the MTA negotiation process.

2.   Mutually Agreed Terms: Documentation of the achievement of agreement terms recognizing at least three ownership groups: government, local/indigenous communities, and private.
Recommended: Submission of signed MTA signifies terms have been agreed.

3.   Prior Informed Consent: Documentation of the provision of sufficient information to reach a reasoned decision.
Recommended: Two standards be established, one for research and the other for commercial use, with higher information requirements for research use involving a partial donation.

4.   Benefit Sharing: Division of benefits among the government and other national stake holders.
Recommended: Governments claim as a right the opportunity to negotiate certain national benefits including samples, national royalty-free use, employment of national scientists, as well as other benefits, including monetary compensation, but total government claims should not exceed 50% of the estimated total value of the resource when other stakeholders are present.

5.   Responsible Authority: National body responsible for interpreting and implementing the law.
Recommended: It is essential for the smooth functioning of access legislation to have a responsible authority with multiple powers, including the authority to declare all requirements have been satisfied, in order for access legislation to operate steadily.

6.   Penalties/Sanctions: Penalties should be clearly stated.
Recommended: Penalties could include cancellation of the access permit, payment of damages, and possible civil and criminal penalties. An arbitration option for resolving disputes should be allowed.

7.   Enforcement: Mechanisms for tracking materials, permitting identification of violations, and enforcement of rights.
Recommended: Require export permits for genetic materials as a requirement for importation into other countries. Downstream, companies could be required to maintain records of the sources of genetic materials.

The following draft access law language incorporates the preceding proposals and is suggested for consideration:

Scope

All living materials, parts, components, and extracts thereof found within the boundaries of *xxxx*, its territories and possessions, but excluding:

- human genetic resources;
- materials acquired prior to the CBD entering into force in December 1993 and held in *ex situ* conditions;
- materials for which the sovereign rights of the government are established by a convention other than the Convention on Biological Diversity;
- materials for which other forms of ownership, such as by IPR, have been established;
- materials deemed to have entered the public domain as indicated by reference, by description, and/or use of a variety name in publications, catalogues, and databases, as described in the implementing regulations.

Access

Permission is required for access, exportation, and use of genetic resources identified under scope. Two categories of use are identified, as follows:

- research use, undertaken for knowledge generation and/or other predominately non-commercial purposes, although recognizing that commercial uses are not precluded and may be identified inadvertently, and for which the suppliers are implicitly being asked to make a contribution in kind to facilitate that knowledge generation; and
- commercial use, undertaken explicitly with the objective of developing marketable products or services.

Mutually agreed terms

Prospective users of controlled genetic resources must document the reaching of mutually agreed terms with the providers of those resources through the filing of a copy of a signed Material Transfer Agreement specifying the terms and conditions of the agreement. The MTA must satisfy other applicable national laws and regulations, including legislation controlling investments and ownership. Parties to those agreements are identified in three groups for controlled materials occurring on:

- public lands;
- traditional/tribal lands; and
- private lands.

Benefit sharing

Benefits are to be shared with the national government and other parties. These parties become involved when the resources occur on communal, traditional or private lands, and are the property of autonomous government institutions, or, when the resources are mobile, are owned by a non-governmental party. Benefits claimed by the government on behalf of the peoples of *xxxx* may include:

- samples of materials collected and related data and information, provided that the Competent Authority establishes procedures for evaluating the user's requests for the need for confidentiality;
- royalty-free domestic use if mutually acceptable;
- involvement of *xxxx*'s scientists and/or technicians in collection and taxonomy, including training as needed and appropriate,
- negotiation of additional benefits payable on successful commercialization including possible financial returns or returns in kind such as capacity building, and transfer of structures and/or equipment. In cases where a benefit sharing agreement is also negotiated with the owners of the materials, the share received by the government, as determined by an impartial comparison of the terms of the agreements, shall not exceed *xx* percent of the total benefits.

Non-governmental owners are authorized to negotiate independently a benefit sharing agreement with the principals (users) which may specify monetary and/or non-monetary compensation, prior to and/or on successful commercialization, provided no other provisions of this law are violated.

Responsible authority

The Government shall establish a Competent Authority for the interpretation and administration of this access legislation. The Authority shall have the following duties and responsibilities, plus others identified and delegated by the Government:

- serve as the national contact point for accessing genetic resources;
- establish the requirements for determining which agreements are research and which commercial and, on request, certify specific requests as one of the two;
- process applications;
- identify and make available national legislation and/or procedures describing the ownership of genetic resources occurring on public, private, and traditional lands;
- on request, identify the legal owners of materials located in specified areas and identify the parties to be involved in the negotiations;

- establish PIC requirements;
- certify the fulfilment of the PIC requirements;
- review submitted MTAs and designate the transactions as research or commercial.

The Authority shall be provided with manpower and financial resources as required to discharge these responsibilities according to the established time table. The Authority may be granted the authority by the Government to assess an applicant fee sufficient to cover the operating expenses.

Penalties/sanctions

Entities which do not fulfil the terms of agreements established under this Act are liable for cancellation of the access permit, payment of damages, and possible civil and criminal penalties. All non-criminal penalties, on mutual agreement, may be brought to binding arbitration according to the procedures of *xxxx*.

Entities which export and/or use genetic resources covered by this Act without complying with it are liable for civil and or criminal penalties including but not limited to fines as specified in *xxxx*.

Needs and approaches

- Identify a central organization for providing national assistance for countries/communities considering access legislation, including the sharing of experiences and perspectives from governments, communities, and private sector representatives.
- These needs can be served by the 'Facilitator', in part by using the Clearing-house Mechanism.

Valuation and Equity (Articles 8, 15, 16, and 19)

Of the multiple definitions and measures of value, two are of particular significance here, *use* (direct and indirect) and *non-use* (option, existence, and bequest) values. A second distinction can be drawn between *public* (social) and *private* value. Public value typically exceeds private value as individuals cannot capture all of the value of a product or service.

A review of the literature of estimates for genetic resources values, either as individual items or in combination as an ecosystem, indicates the total economic value of a habitat can be very large. A figure of $30 billion has been proposed for the Brazilian Amazon. Much of this value, however, is of the non-use type. Direct use values are far smaller, sometimes a few hundred dollars a sample. Additional work, including methodological advances, are needed to provide more accurate estimates.

While direct use values, referred to as *willingness to pay*, are often small, the position taken here is that they are most indicative of actual cash payments for genetic resources. When relatively low willingness to pay values are estimated, great care must be taken in the use of available monies so the incentive effect for the conservation of genetic resources is maximized. Moreover, low estimated values indicate that exchange transactions for genetic resources must be efficient so as not to absorb the value of the materials.

Determining equity requires understanding what the concept means in practice, in the context of a particular agreement. Several approaches are presented, differing on whether interpersonal welfare comparisons ('the rich can afford to pay more') need be made. In general, systems using interpersonal comparisons are more intuitively appealing, but prove difficult for achieving a consensus. Non-comparison systems provide a more specific approach, but with the notable limitation of excluding wealth from the consideration of equity. This means, among other things, there can be no decisions based on the premise that the wealthy firm 'should' pay a higher price because it 'can afford to'.

Here, the *marginal value* (a non-comparison) approach to equity is endorsed as promising contributors are rewarded according to the market value of their contributions. The requisite conditions for this approach to advance equity are, however, significant; in addition to ignoring differences in wealth and assuming that no unpaid components exist in the economy (such as lack of pollution), the marginal value approach requires the market be perfectly competitive. It will be a major task of governments to maximize the achievement of these requirements, including the maintenance of an open economy.

Needs and approaches

- Expand research work on the valuation of genetic resources, especially the valuation of individual materials, including methodological improvements.
- Discuss in the proper fora, what is meant by 'fair' and 'equitable' so as to provide a common understanding for future COP discussions. Different perspectives likely contribute to disagreements which hampers the search for practical action. Here the limited 'marginal value' approach is recommended as being relatively value free and identifying a number of steps to be taken by national governments and international bodies to enhance equity.

Prior Informed Consent (Article 15)

PIC for access to genetic resources presents broad and more complex issues, including its justification. The underlying rationale is personal autonomy, a Western concept based on the eighteenth century philosophy of the Enlightenment. While Western in its expression, autonomy can be seen as a component of integrity, integrity of the person and personal integrity, which incorporates personal and societal harmony. Hence, in this broader perspective, the role of informed consent can be established. Yet it remains that those providing the information are placed in a special, elevated context of obligation to those from whom the consent is being sought.

Two cases can be identified, the first is where the consent being sought is for research of no clear direct benefit to the consentor. It is in this context that medical PIC has evolved over more than a century. Consentors, for example, may wish to know the objectives of the research, which they may support or not according to their values and beliefs. The possibility of generating commercial products from the research would be one bit of pertinent information to be revealed through PIC.

The second concerning strictly commercial access agreements, however, has a conceptual basis quite different as the supplier is seeking specific (monetary) benefits. Expecting contractors (buyers) to provide information which could weaken their negotiating position in respect to sellers is a high standard to establish indeed. Law, however, does routinely establish disclosure and information standards, which contractors are advised to read carefully before signing.

As a first guiding principle, country PIC requirements should establish different standards/requirements for commercial and research agreements. The provision of information needed by the seller to assist in striking a favourable commercial agreement should not be the responsibility of the buyer/contractor. Rather, sellers need to recognize market intelligence as being their own responsibility. In practice, some sellers will need assistance, information, and capacity building to strengthen their negotiating position, at least in the initial stages.

For primarily research agreements, far more demanding PIC requirements can be justified. Suppliers are being asked to make a contribution to the public good, and should not be expected to take on the cost of informing themselves of the possible outcomes. A neutral party should also be involved, possibly as identified under national legislation, to be financed by the would-be contractor. The shared information should be adequate for the supplier to predict the possible consequences of the agreement. Comprehension can be verified by asking the supplier to restate the information in his/her own words and explain the decision reached. When conditions require, this may be done verbally (and preserved on tape as verification) in the second of a 'two part' consent form.

The following suggested language encapsulates these concepts:

Prior informed consent

Prior informed consent requirements are related to the intended use of the materials as primarily commercial or research.

Commercial use is determined by the project description and agreement represented in the signed Material Transfer Agreement. Prospective users must disclose the intended use of the products produced, and any known risks associated with the collection of the materials or production/use of the products. Sellers are responsible for negotiating an agreement. Consent is indicated by the sellers' representative(s) signing a disclosure form, the original of which must be submitted to the Competent Authority.

Research use is that undertaken for the generation of knowledge for which the supplier(s) are not expected to benefit directly. Because the suppliers are being asked in effect to make a public donation, the researchers are considered to have a higher information obligation than for commercial agreements. In particular, researchers are expected to provide the following information:

- purposes of the research and anticipated benefits/beneficiaries;
- possible commercial applications of the research;
- any risks to the suppliers; and
- alternative approaches,

among other information required by the Competent Authority and listed in the Regulations to this legislation. Suppliers are responsible for establishing in the MTA the: (i) permissible conditions of commercialization and (ii) any remuneration or other benefits owing upon commercialization.

Informed consent is certifiable by either: (i) the signing of the disclosure form in the case of commercial agreements or (ii) the signing or video-taping of the presentation of information and consent by suppliers in the case of research agreements. The decision of which medium to use, paper copy or video, is to be made by and announced by the Competent Authority based on the Authority's assessment of the situation, and particularly the literacy level in the native language of the suppliers. The Authority may also require the consent process further include a restatement by the suppliers in their own words of the information provided and explanation of the decision reached. The Authority shall then determine if the suppliers' level of comprehension is sufficient to provide informed consent, and if not shall specify what additional steps will be required.

Needs and approaches

- The concept of prior informed consent needs to be operationalized; as a concept only, it contributes to misunderstanding and dissatisfaction.

- PIC requirements should be established in national access legislation in sufficient detail to make requirements and responsibilities clear. A Competent Authority needs to be identified with the power to rule when PIC requirements have, or have not, been satisfied.
- Recommended here is a two-tiered PIC system for materials use for commercial and research purposes, with the information requirements high in the latter case.

Materials Held *Ex Situ* (Article 9)

Until very recent times, genetic materials have been treated as a common heritage, a principle which underlay the creation of the world's significant *ex situ* collection of plant genetic resources for food and agriculture. Treated as a common heritage with no charges or limitations imposed, access and use have been excellent for the development of food crops, less so for commodities and industrial crops. Evident strains on this system began in the 1980s when the supplying countries became dissatisfied with a common heritage system providing no direct compensation.

Ex situ materials receive particular attention under the CBD because those collected prior to 1994 (which is to say, the vast bulk) are not covered by the terms of the Convention. Yet a resolution of the status of these materials is mandated in the Nairobi Final Act and is clearly a top priority for the Parties – for scientific, humanitarian, and financial reasons.

As part of the access and use process, a significant (and, to the outside observer) complex institutional mechanism has been established by the FAO. That system produced the concept of Farmers' Rights and the Global Plan of Action. Those mechanisms are presently being harmonized with the CBD. There have been broad agreements on the details and the general objectives, the partitioning of materials into open (designated) and restricted access materials. These efforts, one after another, have however been stymied by the question of who pays whom, and for what. The current version, described in the 'MUSE' report, suggests a number of alternative yet standardized approaches, but does not resolve the payment conundrum. The cost figures, however, are convincing for why a general bilateral system will not work well; costs are simply too high.

A significant aspect of the FAO activities has been the identification of Farmers' Rights, essentially a fund for compensating traditional farmers for developing and conserving germplasm for food and agriculture. This has been an extremely contentious issue over the past 14 years, with little progress made towards generating the funds. Depending on exactly how the funds would be used, Farmers' Rights are more a matter of equity than incentives for future activities, including conservation.

One way ahead is to *separate access to the material from access to the information regarding it.* The material itself is a common heritage resource, including 'passport data' identifying where collected. The genetic material, however, has but modest value with such limited characterization. In the absence of other arrangements, passport and all other available information is currently distributed without charge with most accessions. One consequence is that most entities have no incentive to invest in screening, so much useful material is never identified and used. Presently, an investment is made in screening when no known material has, say, a desired disease resistance. Private entities have little incentive to provide the requisite investment in growouts and/or testing or if they do, they protect the results with IPR and/or secrecy. A system which facilitates payments for information would provide incentives for national programmes to add value through screening trials.

The issue none the less remains – what incentive exists for national programmes to donate their genetic resources for public use while selling associated information. Programmes could after all sell both the information *and* materials. Voluntarily limiting sales to information could require an inducement involving some external (e.g. non-market donor) support, as converting common property to private property for which payment may be collected is complex and costly in transaction costs.

Needs and approaches

- Complete expeditiously the harmonization of the FAO activities regarding genetic resources with the CBD. As major new and additional funding is not evident at this time, plans which require large funding increases are not likely to be viable.
- Moderate the focus on Farmers' Rights as past discussions have increased divisions with few benefits evident. If Farmers' Rights are to be pursued, then some specific indications of how the funds would be used would focus the discussions in a positive way. A name change might help bypass entrenched positions; 'Genetic Resources Foundation' is suggested.
- Operationalize the proposed MUSE system to assure the continued open availability of genetic materials.
- Rather than supplier countries attempting to claim payments for genetic materials themselves, with the inherent complexities and costs, a system for donating the materials but selling characterization data is proposed. Such a system would add value, not only redistribute monies. Donors could assist natural programmes in establishing the screening programmes and the SINGER project in marketing the information.

Property Rights for Indigenous/Local Knowledge (Articles 8 and 18)

The failure of existing property rights systems to provide protection and benefits to local and indigenous communities is one of the more contentious issues of the Convention. Conceptually, IPR protection for technologies in the areas of genetic resources and landraces fits smoothly within the historical development of IPR legislation. Practically speaking, however, current legislation is applicable to improved plant varieties but is not readily suited to landraces and the like, even though they may be technically protectable. For other materials, patents are usable, but the costs of documenting and preparing an application make patents a prohibitive approach for the great bulk of materials of uncertain use and value. In cases when patents are applicable, a financing system would be required to assist local and indigenous communities in protecting their inventions.

The same general conclusions stated above apply when the consideration is extended to 'non-traditional' property rights systems like codes of conduct, appellations of origin, and systems for folklore. One approach receiving much current attention is the patent requirement of source documentation, but this would apply, at best, only to the small amount of material which is patented and leaves the question of derivatives ownership unresolved. Alternatively, language could be included unilaterally in national access legislation, or could operate on a reciprocal basis. Or it could be considered as a protocol under the Convention. Suggested enabling language follows:

> Importation of genetic materials covered by the Act, their parts and components, must be accompanied by a certificate from the Competent Authority in the supplying country attesting the materials were acquired in accordance with national access laws and regulations. Third party possessors of genetic resources must supply a copy of the original certificate of compliance. If the materials to be imported are excluded from the exporting country's access legislation, or the country has no such requirements in place at the time of the transfer, the applicant must provide a sworn statement to that effect, subject to criminal penalties for false or misleading information.

Enforcement mechanisms depend on the means of implementation. If done through national access laws, two approaches are open: (i) refusal to deal in the future, and (ii) suing for contract violation in the user's country. Employing the latter case requires a consideration of the national system of laws where any disagreement will be litigated, including any reciprocal agreement among countries.

One solution for protecting this knowledge lies in the use of material transfer agreements, bolstered by the necessary requirements in access legislation. The use of MTAs, however, requires secrecy which is difficult, limiting to the exchange of knowledge, and may be contrary to cultural norms.

The alternative is a *sui generis* system. The complexity of that undertaking should not be underestimated as it involves protecting unembodied knowledge. Here is proposed a system which acts to 'reserve' knowledge by identifying it in printed form and requiring permission from the owner(s) when so-identified knowledge is commercialized. The concept is distinct in separating ownership of the genetic material (which is often ambiguous) from the knowledge (which is far clearer). Further development of this and other approaches is needed, including a larger participation by the private sector, the major user of the materials.

Suggested language for a *sui generis* system follows:

Clause for inclusion in access legislation

This research agreement provides rights for research purposes only. Any commercial use of the collected materials, or the compounds or structures identified directly or indirectly therefrom, is subject to a separate commercial use agreement. Mutually acceptable agreements must be negotiated with:
(a) the owner(s) of the genetic resources as identified in this research agreement, and
(b) for genetic materials listed in the *Registry of Traditional Uses* included in the Regulations for this legislation, the community representative(s) identified therein.

Regulations

Registry: *The Registry of Traditional Uses* shall consist of references to published listings of genetic resources uses for agricultural, medicinal and other purposes. Listings shall include:
1. Descriptions of use(s), to include all materials incorporated and any preparatory practices. In case of multiple listings of the same genetic materials, the first published listing shall have priority.
2. Scientific names, descriptions and/or photos of genetic resources, employed in the above uses in sufficient detail such that the materials can be identified by knowledgeable individuals.
3. Identification of community group representative(s) to be contacted for conducting commercialization negotiations. Inclusion of these names signifies that the identified communities have been previously informed of the possibility of commercial use.
4. Locally published, out of print, or otherwise difficult-to-acquire listings shall be deposited in the national library and there made reasonably available, along with photostat service availability at prevailing government rates.

Duration

Community claims to knowledge shall be limited to 50 years from first publication in the Registry of Traditional Uses, after which time the knowledge shall be considered to be part of the public domain.

Dispute settlement

Claims of indigenous use shall be subject to documentation and verification at the request and expense of the collector. Disputes shall be heard in the [nationally named] court. Parties may select to have disputes settled by binding arbitration.

Claims of indigenous and local use are subject to judicial review, or binding arbitration.

Needs and approaches

- IPR systems are not well suited to protecting traditional/indigenous knowledge. In the limited cases where they are applicable, funding may be needed by local communities to complete the protection process. National and/or international funds, possibly with a contribution by the CBD Financial Mechanism, are needed.
- Proposals to date for *sui generis* systems are not sufficiently developed to be operational, or have other limitations. Or, the requirement that the source of materials be identified on patent applications could apply to a limited number of cases where patents are used, and would at best be slow to implement. A requirement included in national access legislation would accomplish much of the same objective, but would necessitate a consideration of where the MTA would be adjudicated.
- In recognition of the problem of many local and indigenous communities in establishing ownership to genetic materials *per se*, an approach for a *sui generis* system is proposed which separates ownership of materials from ownership of knowledge, and protects the latter.

Technology Transfer – the Roles of Government (Article 16)

Under the CBD, developed countries in particular have made general commitments to assist in the transfer of technologies relevant to biodiversity. In practice, much of that technology is privately owned, leaving the governments' roles ambiguous.

More specifically, a three-tiered ranking of importance of roles in technology transfer arrangements can be identified, as follows:

- arrangements between buyers and sellers;
- policies of recipient countries in such areas as regulations, ownership, taxes, and IPR; and
- policies of supplying countries.

Hence, it is not surprising that the planned activities of developed countries as facilitators of technology transfer are little developed. What, though, might those roles effectively be?

Virtually all knowledgeable observers, from inside and outside the CBD, governments, companies, and the NGO community, recognize the need for *capacity building* and *funding*. These are traditional public sector roles which are no less appropriate in technology transfer. Funding relates primarily to making available expensive technologies. Capacity building, for its part, is seen as required for a range of functions from general technical/scientific training to local adaptation to negotiation skills.

Beyond these general public sector roles of capacity building and facilitation, some more specific critical functions can be identified as follows:

Risk shifting. Including financial risks and liability.

Concessional terms. Payment in national, non-convertible currencies which may be used domestically for national biodiversity conservation.

Reducing management requirements. One of a corporation's most scarce commodities is time, yet biotechnology transfer, being still a novel practice, is inordinately time consuming. This means that, as a practical matter, donations and arrangements for small markets simply rarely get completed. A major contribution would be the standardization of agreements, the creation of a model contract. This may not be used in its entirety but can function as the basis for negotiations with the longer term objective of routinizing exchange arrangements. Developing these agreements will be an educational as well as a legal/commercial process.

Needs and approaches

- For a variety of reasons, including the private ownership of many technologies relevant to the achievement of the convention's objectives, the role of developed country governments in facilitating technology transfer will largely be limited to capacity building and funding.
- Identified broader activities, where appropriate, include risk shifting and reducing management requirements.

Technology Transfer – the Roles of Intellectual Property Rights (Article 16)

The function and roles of IPR have been among the most contentious issues under the CBD since the Rio Conference. This is partly due to the issue of rights for traditional/indigenous knowledge, and partly to a misunderstanding of the IPR functions in technology transfer.

The Convention, by emphasizing transfer, implicitly focuses on existing technologies so that it is the *access* role of IPR which would be more relevant. Regarding access, several competing forces exist depending on the specific technology: the fear of losing control of a technology versus the impossibility of maintaining long-term control; and the potential for using secrecy/physical control as a protective mechanism. Overall, as theory and experience indicate, limited IPR probably means delayed access with other, if less visible, costs. IPR systems have readily observable costs, royalty payments being the most obvious, but the costs of the absence of protection in terms of denied or delayed access must be accessed on a case-by-case basis. Yet, information on the balance of these costs is limited. The CBD Secretariat has taken an initial step by requesting the submission of case studies on the functioning of IPR in technology transfer.

While a significant level of dissenting opinion does and will continue to exist, strengthened IPR seems the direction of the future. This conclusion coincides with national commitments under TRIPs, as well as national needs for access to technology. Indeed, many countries, even long-time severe critics like Brazil and India, are strengthening IPR legislation, in a large measure for domestic reasons. The issue then arises of what forms of law and system are appropriate for the existing diversity of nations? Clearly, the level of detailed knowledge about the practical operation of these systems across a broad range of countries is limited.

Overall, there is an apparent need for assisting countries in the consideration of alternatives for the extension/modification of IPR. Presently, WIPO provides training programmes for future examiners, but what is now needed is broader and possibly more controversial as questions need to be asked about the roles of IPR at differing levels of economic development. Just which body or bodies would be responsible is not clear. Certainly, multinational firms using IPR and major IPR offices, including those in the USA, Japan, and EPO, should participate, but additional experiences and perspectives are needed as well. A coordinating body for providing that support needs to be identified.

Needs and approaches

- IPR is neither a necessary nor sufficient requirement for technology transfer, but available evidence indicates adequate IPR protection does facilitate access to technologies, particularly readily copied ones like biotechnology.
- In many countries the patentability of plants and animals, and even gene sequences, remains ambiguous. That situation complicates decisions on technology transfer and needs clarification, possibly by the placement of test patent applications in key countries.

References

Abbott, A. (1993) Monoglot filing urged for European Patents. *Nature* 364, 3.

ACTS (African Centre for Technology Studies) (1993) Convention on Biological Diversity: National Interests and Global Imperatives. Nairobi.

Anderfelt, U. (1971) *International Patent Legislation and Developing Countries.* Martinus Nijhoff, The Hague.

Anderson, M. (1989) International Technology Transfer in Agriculture. US Department of Agriculture, Economic Research Service, Agricultural Information, Bulletin 571, Washington, DC, August.

Andes Pharmaceuticals (1995) Company Profile. Washington, DC, October.

Artuso, A. (1994) Economic Analysis of Biodiversity as a Source of Pharmaceuticals. Paper presented at the PAHO/IICA Conference on Biodiversity, Biotechnology and Sustainable Development, San José, 12–14 April.

Asebey, E.J. and Kempenaar, J.D. (1995) Biodiversity prospecting: fulfilling the mandate of the biodiversity convention. *Vanderbilt Journal of Transnational Law* 28(Oct.), 703–754.

Aylward, B.A. (1993) A case study of pharmaceutical prospecting. In: Aylward, B.A., Echeverria, J., Fendt, L. and Barbier, E.B. (eds) *The Economic Value of Species Information and Its Role in Biodiversity Conservation: Case Studies of Costa Rica's National Biodiversity Institute and Pharmaceutical Prospecting.* Report to Swedish International Development Authority, Part III.

Bagnara, D., Bagnara, G. and Sontaniello, V. (1996) Role and Value of International Germplasm Collections in Italian Durum Wheat Breeding Programs. Paper presented at CEIS-Tor Vergata Symposium on the Economics of Valuation and Conservation of Genetic Resources for Agriculture, Rome, May 13–15.

Barbault, R. and Sastrapradja, S. (lead authors) (1995) Generation, maintenance and loss of biodiversity. In: Heywood, V.H. (exec. ed.) *Global Biodiversity Assessment.* Cambridge University Press, Cambridge (published for UNEP), Chap. 4.

Barton, J.H. and Siebeck, W.E. (1994) Material transfer agreements in genetic resources exchange – the case of the International Agricultural Research Centres. IPGRI, Issues in Genetic Resources No. 1, Rome, May.

Barton, J.H., Dellenbach, R.B. and Kuruk, P. (1989) Toward a theory of technology licensing. *Stanford Journal of International Law* 25, 195–228.

Barzel, Y. (1989) *Economic Analysis of Property Rights*. Cambridge University Press, Cambridge.

Belson, N.A. (1996) The Biodiversity Convention and the private sector. *Diversity* 12, 6–7.

Bent, S.A., Schwaab, R.L., Conlin, D.G. and Jeffrey, D.D. (1987) *Intellectual Property Rights in Biotechnology Worldwide*. Stockton Press, New York.

Bentley, J.W. (1993) What farmers don't know. *Ceres* 14(May–June), 42–45.

Bhatia, S. and Kothari, A. (1996) Community Register for Documenting Local Community Uses of Biological Diversity. Prepared for the Stakeholder Workshop on Implementation of Articles 15 and 16 of the Convention on Biological Diversity by the European Union, London, Feb. 12–13.

Biotech Reporter (1995) Jeremy Rifkin Attacks WR Grace's Neem Patent. October, pp. 1 & 3.

Blumenthal, D., Causino, N., Campbell, E. and Louis, K.S. (1996) Relationships between academic institutions and industry in the life sciences – an industry survey. *New England Journal of Medicine* 334(Feb. 8), 368–373.

Brim, C. (1987) Plant breeding and biotechnology in the United States of America: changing needs for protection of plant varieties. In: *Proceedings of the World Intellectual Property Organization Symposium on the Protection of Biotechnological Inventions*, June 4–5, Ithaca, New York.

Brockway, L.H. (1988) Plant science and colonial expansion: the botanical chess game. In: Kloppenburg, J.R. Jr. (ed.) *Seeds and Sovereignty: The Use and Control of Plant Genetic Resources*. Duke University Press, Durham, North Carolina, Chap. 2.

Brush, S.B. (1993) Indigenous knowledge of biological resources and intellectual property rights: the role of anthropology. *American Anthropologist* 95, 653–670.

Brush, S.B. (1996a) Whose knowledge, whose genes, whose rights? In: Brush, S.B. and Stabinsky, D. (eds) *Valuing Local Knowledge: Indigenous Peoples and Intellectual Property Rights*. Island Press, Washington, DC, Chap. 1.

Brush, S.B. (1996b) Is common heritage outmoded? In: Brush, S.B. and Stabinsky, D. (eds) *Valuing Indigenous Knowledge: Indigenous Peoples and Intellectual Property Rights*. Island Press, Washington, DC, Chap. 7.

Brush, S.B. and Meng, E. (1996) Farmers' Valuation and Conservation of Crop Genetic Resources. Paper presented at CEIS-Tor Vergata Symposium on the Economics of Valuation and Conservation of Genetic Resources for Agriculture, Rome, May 13–15.

Burrows, B. (1997) Second thoughts about US patent #4,438,032. *Bulletin of Medical Ethics* 124(Jan), 11–14.

Busch, L., Lacey, W.B., Burkhardt, J. and Lacey, L.R. (1991) *Plants, Power and Profit – Social, Economic and Ethical Consequences of the New Biotechnologies*. Basil Blackwell, Oxford, UK.

Butler, B. and Pistorius, R. (1996) How farmers' rights can be used to adapt plant breeders' rights. *Biotechnology and Development Monitor* 28(Sept.), 7–11.

Butler, L.J. and Marion, B.W. (1985) *The Impacts of Patent Protection on the US Seed Industry and Public Plant Breeding*. University of Wisconsin, Agricultural Experimental Station, NC Project 117, Monograph 16, Sept.

Chapman, A.R. (1994) Human rights implications of indigenous peoples' intellectual property rights. In: Greaves, T. (ed.) *Intellectual Property Rights for*

Indigenous Peoples. Society for Applied Anthropology, Oklahoma City, Oklahama, Chap. 14.

Chase, L.C., Lee, D.R., Schulze, W.D. and Anderson, D. (1998) 'Ecotourism demand and differential pricing of National Park access in Costa Rica'. *Land Economics* (in press).

Claffey, B.A. (1981) Patenting life forms: issues surrounding the Plant Variety Protection Act. *Southern Journal of Agricultural Economics* 13, 29–37.

Coe, R.N. (1994) Keeping trade secrets secret. *Journal of the Patent and Trademark Office Society* 76, 833–840.

Cooper, D., Engels, J. and Frison, E. (1994) A multilateral system for plant genetic resources: imperatives, achievements and challenges. Issues in Genetic Resources No. 2, May, IPGRI, Rome.

Cooper, P. (1984) *Plant Breeders' Rights: Some Economic Considerations.* Economic Working Papers, Agriculture Canada, Ottawa.

Correa, C.M. (1994) *Sovereign and Property Rights Over Plant Genetic Resources.* FAO Background Study Paper No. 2, Nov., Rome.

Crespi, R.S. (1992) Patents and plant variety rights: is there an interface problem? *International Review of Industrial Property and Copyright Law* 23, 168–184.

Cromwell, E., Friis-Hansen, E. and Turner, M. (1992) *The Seed Sector in Developing Countries: A Framework for Performance Analysis.* Working Paper 65, Overseas Development Institute, London.

Damania, A.B. (1996) Swaminathan foundation holds technical consultation to develop framework for farmers' rights in India. *Diversity* 12, 6–8.

de Freitas, T.C. and de Carvalho, P.A.O. (1996) Brazil Overhauls IP Law Sept./Oct., p. 20.

de Janvry, A. and Dethier, J.-J. (1985) Technology Innovation in Agriculture: The Political Economy of Its Rate and Bias. CGIAR Study Paper No. 1, Washington, DC.

DeGregori, T.R. (1985) *A Theory of Technology: Continuity and Change in Human Development.* Iowa State University Press, Ames, Iowa.

Deolalikar, A.B. and Evenson, R.E. (1990) Private inventive activity in Indian manufacturing: its extent and determinants. In: Evenson, R.E. and Rains, G. (eds) *Science and Technology: Lessons for Development Policy.* Westview Press, Boulder, Colorado, Chap. 10.

Diversity (1996a) 'Conference Fatigue' takes toll on FAO Commission as process to harmonize undertaking and biodiversity convention begins. 12, 8.

Diversity (1996b) Indigenous peoples' concerns dominate COP-3 deliberations. 12, 27–28.

Downes, D., Laird, S.A., Klein, C. and Carney, B.K. (1993) Biodiversity prospecting contract. In: Reid, W.V. *et al.* (eds) *Biodiversity Prospecting: Using Genetic Resources for Sustainable Development.* World Resources Institute, Washington, DC, Annex 2, pp. 255–287.

Duesing, J. (1992) The convention on biological diversity – its impact on biotechnology research. *Agro-Industry Hi-Tech* 3, 19–23.

Duffey, W.H. (1989) The marvelous gifts of biotech: will they be nourished or stifled by our international patent laws? In: *Proceedings of the World Intellectual Property Organization Symposium on the Protection of Biotechnological Inventions,* 4–5 June, Ithaca, NY, pp. 27–36.

Earth Negotiations Bulletin (1996) A brief analysis of COP-3. International Institute for Sustainable Development (http://www.iisd.ca/linkages/).

Economist (1996a) The buying and selling of Brazil, Inc. 9–15 Nov., 83–84.

Economist (1996b) Pharmaceuticals in India: best of both worlds. 7 Dec., 60, 63.

Economist (1996c) Limited imagination, 28 Sept., 80 and 85.

Engels, J.M.M. (ed.) (1995) *In situ* conservation and sustainable use of plant genetic resources for food and agriculture in developing countries. Report of a DSE/ATSAF/IPGRI workshop, 2–4 May, Bonn-Roettgen, Germany, IPGRI and DSE, Rome and Feldafing.

Evenson, R.E. (1988) Technological opportunities and international technology transfer in agriculture. In: Antonelli, G. and Quadrio-Curzio, A. (eds) *The Agro-Technological System Towards 2000*. Elsevier Science Publishers, New York, Chap. 7.

Evenson, R.E. (1990) Intellectual property rights, R&D inventions, technology purchase, and piracy in economics development: an international comparative study. In: Evenson, R.E. and Rains, G. (eds) *Science and Technology: Lessons for Development Policy*. Westview Press, Boulder, Colorado, Chap. 14.

Evenson, R.E. (1994) *The Value of Genetic Resources for Crop Improvement: Evidence from Rice, Mimeo*. Economic Growth Center, Yale University.

Evenson, R.E., Evenson, D.D. and Putnam, J.D. (1987) Private sector agricultural invention in developing countries. In: Ruttan, V.W. and Pray, C.E. (eds) *Policy for Agricultural Research*. Westview Press, Boulder, Colorado, Chap. 19.

Eyzaguirre, P. and Iwanaga, M. (eds) (1996) *Participatory Plant Breeding*. IPGRI, Rome.

FAO (1993) International Code of Conduct for Plant Germplasm Collecting and Transfer. CPGR/93/8, Commission on Plant Genetic Resources, FAO Conference 27th November Session, Rome.

FAO (1996a) Draft Report of the State of the World's Plant Genetic Resources. D/W0986, 1023, 1055, April, Rome.

FAO (1996b) Report: International Technical Conference on Plant Genetic Resources. ITCPGR/96/REP, Rome.

FAO, Commission on Plant Genetic Resources (1995) Revision of the International Undertaking on Plant Genetic Resources – Analysis of Some Technical, Economic and Legal Aspects for Consideration in Stage II: Access to Plant Genetic Resources and Farmers' Rights. CPGR-6/95/8 Supp., June, Rome.

Finesilver, J.M. (1995) Biodiversity Prospecting: Prospects and Realities. In: Zakri, A.H. (ed.) *Prospects in Biodiversity Prospecting*. Genetics Society of Malaysia, Selangor, Malaysia, Chap. 2, pp. 21–58.

Fitzgerald, D. (1990) *The Business of Breeding: Hybrid Corn in Illinois, 1890–1940*. Cornell University Press, Ithaca, New York.

Fowler, C. (1995) Biotechnology, patents and the Third World. In: Shiva, V. and Maser, I. (eds) *Biopolitics: A Feminist and Ecological Reader on Biotechnology*. 2nd Books, London, Chap. 13.

Gaber, P.M. (1989) Tulipmania. *Journal of Political Economy*, 97(June), 535–560.

Gadgil, M. and Devasia, P. (1995) Intellectual property rights and biological resources: specifying geographical origins and prior knowledge of uses. *Current Science* 69(Oct.), 637–639.

Glowka, L., Burhenne-Guilmin, F. and Synge, H. (1994) *A Guide to the Convention on Biological Diversity.* Policy and Law Paper No. 30. IUCN Environmental Law Center, Gland, 161 pp.

Gollin, M.A. (1993) An intellectual property rights framework for biodiversity prospecting. In: Reid, W.V. *et al.* (eds) *Biodiversity Prospecting: Using Genetic Resources for Sustainable Development.* World Resources Institute, Washington, DC, Chap. VI, pp. 159–197.

Graham, S. (1956) Class and conservatism in the adoption of innovations. *Human Relations* 9, 91–100.

Greaves, T. (1994) IPR, a current survey. In: Greaves, T. (ed.) *Indigenous Property Rights for Indigenous Peoples: A Source Book.* Society for Applied Anthropology, Oklahoma City, Oklahoma, Chap. 1.

Grossman, R. (1988) Equalizing the flow: institutional restructuring of germplasm exchange. In: Kloppenburg, J.R. Jr. (ed.) *Seeds and Sovereignty: The Use and Control of Plant Genetic Resources.* Duke University Press, Durham, North Carolina, Chap. 11.

Grubaugh, S.C. (1987) Determinants of direct foreign investment. *Review of Economics and Statistics* 69, 149–152.

Gupta, A.K. (1993) Editorial. *Honey Bee* 4, 2–4.

Gutterman, A.S. (1993) The North–South debate regarding the protection of intellectual property rights. *Wake Forest Law Review* 28, 89–139.

Hansen, L. and Knudson, M. (1996) Property rights protection of reproducible genetic materials. *Review of Agricultural Economics* 18(Sept), 403–414.

Hawksworth, D.L. (1995) The resource base for biodiversity assessment. In: Heywood, V.H. (ed.) *Global Diversity Assessment.* Cambridge University Press, Cambridge, Chap. 8.

Hawksworth, D.L. and Kalin-Arroyo, M.T. (lead authors) (1995) Magnitude and distribution of biodiversity. In: Heywood, V.H. (exec. ed.) *Global Biodiversity Assessment.* Cambridge University Press (published for UNEP), Cambridge, Chap. 3.

Heberlein, T.A. and Bishop, R.C. (1986) Assessing the validity of contingent valuation: three field experiments. *Science of the Total Environment* 56, 99–107.

Hecht, S. and Cockburn, A. (1989) *The Fate of the Forest: Developers, Destroyers and Defenders of the Amazon.* Verso, New York, 266 pp.

Heureux, C.J., Kone, S. and Walla, K. (1992) Agribusiness and Public Sector Collaboration in Technology Development and Use in Sub-Saharan Africa: The Case of Crop Protection in Cameroon, USAID AFR/ARTS/FARA, May/June, Abt Associates, Bethesda, Maryland.

Hughes, J. (1988) The philosophy of intellectual property. *Georgetown Law Journal* 77, 287–366.

Hunter, D., Sommer, J. and Vaughan, S. (1994) *Concepts and Principles of International Environmental Law: An Introduction.* Environment and Trade No. 2, UNEP, Nairobi, 51 pp.

Hunter, R.B. (1992) Essentially derived and dependency, some examples. Variety Identification Steering Committee, American Seed Trade Association, Washington, DC, June.

International Institute for Sustainable Development (1996) A Brief Analysis of the Conference. (http://www/iisd.ca/linkages/vol109/094702le.html.)

IPGRI (International Plant Genetic Resources Institute) (1996a) Access to Plant Genetic Resources and the Equitable Sharing of Benefits: A Contribution to the Debate on Systems for the Exchange of Germplasm. Rome, April.

IPGRI (International Plant Genetic Resources Institute) (1996b) Options for Access to Plant Genetic Resources and the Equitable Sharing of Benefits Arising from Their Use. Rome, Dec.

Jaffé, W. and van Wijk, J. (1995) *The Impact of Plant Breeders' Rights in Developing Countries: Debate and Experiences in Argentina, Chile, Columbia, Mexico and Uruguay.* Inter-American Institute for Cooperation on Agriculture, University of Amsterdam.

Jain, H.K. (1982) Plant breeders' rights and genetic resources. *Indian Journal of Genetics* 42, 122.

James C. and Krattiger, A. (1996) *Global Review of the Field Testing and Commercialization of Transgenic Plants: 1986 to 1995 – The First Decade of Crop Biotechnology.* ISAAA Briefs No. 1, Ithaca, New York.

Jayaraman, K.S. (1994) India set to end gene robbery. *Nature* 370(25 August), 587.

Jewkes, J., Sawers, D. and Stillerman, R. (1969) *The Sources of Invention*, 2nd edn. Norton, New York.

Juma, C. and Ojwang, J.B. (eds) (1989) *Innovation and Sovereignty: The Patent Debate in African Development.* ACTS Press, Nairobi.

Juma, C., Mugabe, J. and Ojwang, J.B. (1994) Access to Genetic Resources: Policy and Institutional Issues. African Centre for Technology Studies, Nairobi, Sept.

King, S.R. (1994) Establishing reciprocity: biodiversity, conservation and new models for cooperation between forest-dwelling peoples and the pharmaceutical industry. In: Greaves, T. (ed.) *Intellectual Property Rights for Indigenous Peoples – A Source Book.* Society for Applied Anthropology, Oklahoma City, Oklahoma, pp. 69–82.

Kinnucan, H., Molnar, J.J. and Hatch, U. (1989) Theories of technological change in agriculture with implications for biotechnologies. In: Molnar, J.J. and Kinnucan, H. (eds) *Biotechnology and the New Agricultural Revolution.* Westview Press, Boulder, Colorado, AAAS Selected Symposium 108, Chap. 5.

Kivlin, J.E. (1960) Characteristics of Farm Practices Associated with Rate of Adoption. Ph.D. Thesis, Pennsylvania State University, University Park, PA.

Kloppenburg, J. Jr and Kleinman, D.L. (1987) The plant germplasm controversy. *BioScience* 37(March), 190–198.

Kocken, J. and van Roozendoal, G. (1997) The Neem tree debate. *Biotechnology and Development Monitor* 30(March), 8–10.

Krattiger, A. (1997) *Insect Resistance in Crops: A Case Study of* Bacillus thuringiensis (Bt) *and its Transfer to Developing Countries.* ISAAA Briefs No. 2, Ithaca, New York.

Krueger, A. (1983) *Trade and Employment in Developing Countries, Synthesis and Conclusions.* National Bureau of Economic Research, Washington, DC.

Krummer, K. (1994) *Transboundary Movements of Hazardous Wastes at the Interface of Environment and Trade.* Environment and Trade No. 7, UNEP, Nairobi.

Laird, S.A. (1993) Contracts for biodiversity prospecting. In: Reid, W.V. *et al.* (eds) *Biodiversity Prospecting: Using Genetic Resources for Sustainable Development.* World Resources Institute, Washington, DC, Chap. IV, pp. 99–130.

Laird, S.A. (1996) Access control for genetic resources: the assertion of sovereignty. *Association of Systematics Collections Newsletter* 24(Feb.), 1–3, 6–7, 99–119 and 24(April), 17–23. (Originally published in 1995 by WWF in Gland, Switzerland.)

Lall, S. (1981) *Developing Countries in the International Economy*. Macmillian, London.

Lara, M. del C. and de la Fuenta, J.R. (1990) On informed consent. In: Connor, S.S. and Fuenzalida-Puelma, H.L. (eds) *Bioethics: Issues and Perspectives*. Pan American Health Organisation, Washington, DC, pp. 59–63.

Lawler, A. (1996) Treaty draft raises scientific hackles. *Science* 274 (25 Oct.), 494.

Lesser, W. (1987a) Anticipating UK Plant Variety Patents. *European Intellectual Property Review* 9, 172–177.

Lesser, W. (1987b) The impacts of seed patents. *North Central Journal of Agricultural Economics* 9, 37–48.

Lesser, W. (1987c) Grace periods in first-to-file Countries. *European Intellectual Property Review* 3, 81–85.

Lesser, W. (1991) *Equitable Patent Protection in the Developing World: Issues and Approaches*. Eubios Ethics Institute, Tsukuba, Japan.

Lesser, W. (1994a) Institutional Mechanisms Supporting Trade in Genetic Resources: Issues Under the Biodiversity Convention and GATT/TRIPs. UNEP Environment and Trade No. 4.

Lesser, W. (1994b) *An Approach for Securing Rights to Indigenous Knowledge*. WP No. 15, International Academy of the Environment Biodiversity/Biotechnology Programme, Geneva.

Lesser, W. (1994c) Valuation of plant variety protection certificates. *Review of Agricultural Economics* 16, 231–238.

Lesser, W. (1997) Estimating cost components of alternative systems for distributing genetic materials held *ex situ*. *Association of Systematics Collections Newsletter* 25 (June), 35–40.

Lesser, W.H. and Krattiger, A.F. (1994a) Marketing 'genetic technologies' in South–North and South–South exchanges: the proposed role of a new facilitating organization. In: Krattiger, A.F., McNeely, J.A., Lesser, W.H., Miller, K.R., StHill, Y. and Senanayake, R. (eds) *Widening Perspectives in Biodiversity*. International Academy of the Environment and World Conservation Union, Geneva and Gland, Chap. 5.6.

Lesser, W.H. and Krattiger, A.F. (1994b) The complexities of negotiating terms for germplasm collection. *Diversity* 10, 6–10.

Lesser, W., Krattiger, A.F., Fry, A., Berger, D. and Révéret, J.P. (1995) Implementation of Technology Transfer within the Framework of the Convention on Biological Diversity. International Academy of the Environment, Conches/Geneva, Sept.

Levine, R.J. (1975) The nature and definition of informed consent in various research settings. In: US National Commission for the Protection of Human Subjects of Biomedical and Behavioral Research, *Belmont Report*. Washington, DC, DHEW Pub. No. (OS)78-0014, Chap. 3.

Levingston, R. and Zamora, R. (1983) Medicine trees of the Tropics. *Unasylva* 35, 7–10.

Lindstrom, B. (1996) Biodiversity, ecology, and evolution of hot water organisms in Yellowstone National Park: symposium and issues overview. *Park Science* (Winter), 12, 13, 19.

Macer, D. (1990) *Shaping Genes.* Eubios Ethics Institute, Tsukuba, Japan.

Machlup, F. (1958) *An Economic Review of the Patent System.* Study No. 15, Study of the Subcommittee on Patents, Trademarks and Copyright, Committee on the Judiciary, US Senate.

McKamey, S.H. (1996) Permit Requirements: Tanzania. *Association of Systematics Collections Newsletter* 24(Feb.), 1, 4, 7, 8.

McLeland, L.N. and O'Toole, H.J. (1987) Patent systems in less developed countries: The cases of India and the Andean Pact Countries. *The Journal of Law and Technology* 2, 229–248.

McNeely, J.A. (1988) *Economics and Biological Diversity: Developing and Using Economic Incentives to Conserve Biological Resources.* IUCN, Gland.

McNeely, J. (1995) IUCN and indigenous peoples: how to promote sustainable development. In: Warren, D.M., Slikkerveer, L.J. and Brokensha, D. (eds) *The Cultural Dimensions of Development: Indigenous Knowledge Systems.* Intermediate Technology Publications, London, Chap. 37.

McNeely, J.A., Gadgil, M., Leveque, C., Padoch, C. and Redford, K. (1995) Human influences on biodiversity. In: Heywood, V.H. (exec. ed.) *Global Biodiversity Assessment.* Cambridge University Press (for UNEP), Cambridge, Chap. 11.

Martin, W. and Winters, L.A. (eds) (1995) *The Uruguay Round and the Developing Economics.* Discussion Papers #307, World Bank, Washington, DC.

Meller, M.N. (1997) Costs are killing patent harmonization. *Journal of Patent and Trademark Office Society* 79(March), 211–225.

Milstein, M. (1995a) Yellowstone managers stake a claim on hot-springs microbes. *Science* 270 (13 Oct.), 226.

Milstein, M. (1995b) Park wants money from its microbes. *Billings Gazette* Sept. 20, 1,12. Billings, MT.

Mitchell, R.C. and Carson, R.T. (1990) Using surveys to value public goods: the contingent valuation method. *American Journal of Agricultural Economics* 72, 250.

Mooney, R.P. (1996) Leipzig is what happens to you when you're making other plans. *Diversity* 12, 10–12.

Muchena, O.N. and Vanek, E. (1995) From ecology through economics to ethnoscience: changing perceptions on natural resource management. In: Warren, D.M., Slikkerveer, L.J. and Brokensha, D. (eds) *The Cultural Dimensions of Development: Indigenous Knowledge Systems.* Intermediate Technology Publications, London, Chap. 46.

Myers, N. (1988) Draining the gene pool: the causes, course and consequences of genetic erosion. In: Kloppenburg, J.R. Jr. (ed.) *Seeds and Sovereignty: The Use and Control of Plant Genetic Resources.* Duke University Press, Durham, North Carolina, Chap. 4.

National Research Council (1992) *Neem: A Tree for Solving Global Problems.* National Academy Press, Washington, DC.

Nemoga-Soto, G.R. (1996) The effects of 'Leipzig' on Latin America and the Caribbean. *Biotechnology and Development Monitor* 228(Sept.), 2–5.

Niblett, B. (1995) Arbitrating intellectual property disputes. *Dispute Resolution Journal* 50(1), 64–67.

Nogues, J. (1989) Notes on Patents, Distortions and Development. Mimeo, The World Bank, Washington, DC, Nov. 28.

Nogues, J. (1990) *Patents and Pharmaceutical Drugs: Understanding the Pressures on Developing Countries.* WPS 502, The World Bank, Washington, DC, Sept.

Norgaard, R.B. and Howarth, R.B. (1991) Sustainability and discounting the future. In: Costanza, R. (ed.) *Ecological Economics: The Science and Management of Sustainability.* Columbia University Press, New York, pp. 88–101.

OECD (Organization for Economic Cooperation and Development) (1996) *Intellectual Property, Technology Transfer and Genetic Resources: An OECD Survey of Current Practices and Policies.* OECD, Paris.

Official Journal EPO (1995) Decision of Technical Board of Appeals 3.3.4 dated 21 Feb. 1995. 8, 545–585.

Ollinger, M. and Pope, L. (1995) *Plant Biotechnology: Out of the Laboratory and into the Field.* Agricultural Economics Report No. 697, Economics Research Service, USDA, Washington, DC, April.

Oxford Centre for the Environment, Ethics and Society (1996) Implementing Traditional Resource Rights. Prepared for the Stakeholder Workshop on Implementation of Articles 15 and 16 of the Convention on Biological Diversity by the European Union, London, Feb. 12–13.

Pearce, D. and Moran, D. (1994) *The Economic Value of Biodiversity.* Earthscan Publications (for IUCN), London.

Pellegrino, E.D. (1990) The relationship of autonomy and integrity in medical ethics. In: Connor, S.S. and Fuenzalida-Puelma, H.L. (eds) *Bioethics: Issues and Perspectives.* Pan American Health Organisation, Washington, DC, pp. 8–17.

Perrings, C. (1995) The economic value of biodiversity. In: Heywood, V.H. (exec. ed.) *Global Biodiversity Assessment.* Cambridge University Press (for UNEP), Cambridge, Chap. 12, pp. 822–914.

Persley, G.J. (1990) *Beyond Mendel's Garden: Biotechnology in the Service of Agriculture.* CAB International, Wallingford, UK.

Peters, C.M., Gentry, A.H. and Mendelsohn, R.O. (1989) Valuation of an Amazonian rainforest. *Nature* 339(29 June), 655–656.

Peterson, W. and Hayami, Y. (1977) Technological change in agriculture. In: Martin, L.R. (ed.) *A Survey of Agricultural Economics Literature.* University of Minnesota Press, Minneapolis, Minnesota, Vol. I.

Pistorius, R. (1996) The Leipzig Conference and its backgrounds. *Biotechnology and Development Monitor* 228(Sept.), 4.

Pistorius, R. (1992) Was the US refusal to sign the biodiversity convention necessary? *Biotechnology and Development Monitor* 12(Sept.), 8–9.

Plowman, R.D. (1993) Intellectual property rights in plants – an ARS perspective. *Diversity* 9, 74–76.

Posey, D.A. (1994) International agreements and intellectual property right protection for indigenous peoples. In: Greaves, T. (ed.) *Intellectual Property Rights for Indigenous Peoples – A Source Book.* Society for Applied Anthropology, Oklahoma City, Oklahoma, Chap. 15, pp. 223–251.

Pray, C.E. (1985) *Private Sector Research and Technology Transfer in Asian Agriculture: Report on Phase 1.* Bulletin No. 85–5, Economic Development Center, University of Minnesota, Minneapolis, Minnesota, December.

Pray, C.E. (1986) *Agricultural Research and Technology Transfer by the Private Sector in India.* Report. No. 1, Economic Development Center, University of Minnesota, Minneapolis, Minnesota, October.

Pray, C. and Echeverria, R. (1989) *Private Sector Agricultural Research and Technology Transfer Links in Developing Countries*. Linkages Theme Paper No. 3, ISNAR, The Hague, July.

Primo-Braga, C.A. (1989) The economics of intellectual property rights and the GATT: A view from the South. *Vanderbilt Journal of Transnational Law* 22, 243–264.

Primo-Braga, C.A. (1990) Guidance from economic theory. In: Siebeck, W.E. (ed.) *Strengthening Protection of Intellectual Property in Developing Countries*. Discussion Paper 112, World Bank, Washington, DC, Chap. III.

Primo-Braga, D.A. (1995) Trade-related intellectual property issues: the Uruguay Round Agreement and its economic implications. In: Martin, W. and Winters, L.A. (eds) *The Uruguay Round and Developing Economies*. Discussion Paper 307, World Bank, Washington, DC, Chap. 12.

Principe, P.P. (1991) Valuing the biodiversity of medicinal plants. In: Akerche, O., Heywood, V. and Synge, H. (eds) *The Conservation of Medicinal Plants*. Cambridge University Press, Cambridge, pp. 79–124.

Putterman, D. (1996) Leipzig Conference appraises FAO Global Plan of Action: ownership, equity issues dominate. *Diversity* 12, 7–8.

Rapley, J. (1996) *Understanding Development: Theory and Practice in the Third World*. Lynne Rienner Publishers, Boulder, Colorado.

Rasmussen, J. (1990) The UPOV convention: The concept of variety and technical criteria of distinctness, uniformity and stability, Publication No. 697(E), UPOV, Geneva.

Reichman, J.H. (1993) The TRIPs Component of the GATT's Uruguay Round: Competitive Prospects for Intellectual Property Owners in an Integrated World Market. *Fordham Intellectual Property, Media and Entertainment Law Journal* 4(Summer), 171–281.

Reid, W.V., Laird, S.A., Gamez, R., Sittenfeld, A., Janzen, D.H., Gollin, M.A. and Juma, C. (1993) A new lease on life. In: Reid, W.V. *et al.* (eds) *Biodiversity Prospecting: Using Genetic Resources for Sustainable Development*. World Resources Institute, Washington, DC, Chap. I.

Reiss, A. Jr. (1976) Selected issues in informed consent and confidentiality with special reference to behavioral/social science research/inquiry. In: *US National Commission for the Protection of Human Subjects of Biomedical and Behavioral Research*. Department of Health, Education and Welfare Pub. No. (OS) 78-0014, Washington, DC, Chap. 25.

Rogers, E.M. (1962) *Diffusion of Innovations*. The Free Press, New York.

Root, F.R. (1994) *International Trade and Investment*, 7th edn. South-Western Publishing Co., Cincinnati, Ohio.

Ruttan, V.W. (1984) *Agricultural Research Policy*. University of Minnesota Press, Minneapolis, Minnesota.

Scherer, F.M. (1980) *Industrial Market Structure and Economic Performance*. Rand McNally, Chicago, Illinois.

Shands, H.L. (1994) Germplasm Development and Support for International Genetic Resources. Paper presented at the workshop, Methodologies for recognizing the role of informal innovation in the conservation and utilization of Plant Genetic Resources, Madras, India, 28–31 Jan.

Sheldon, I.M. (1996) Contracting, imperfect information, and the food system. *Review of Agricultural Economics* 18, 7–19.

Sheldon, I.M. and Abbott, P.C. (1996) *Industrial Organization and Trade in the Food Industries.* Westview Press, Boulder, Colorado.

Shiva, V. and Moser, I. (1995) *Biopolitics: A feminist and ecological reader on biotechnology.* Zed Books, London.

Simpson, D. and Sedjo, R.A. (1992) Contracts for transferring rights to indigenous genetic resources. *Resources* 109(Fall), 1–6.

Simpson, R.D. and Sedjo, R.A. (1996) The Value of Genetic Resources for Use in Agricultural Improvement. Paper presented at CEIS-Tor Vergata Symposium on the Economics of Valuation and Conservation of Genetic Resources for Agriculture, Rome, May 13–15.

Simpson, R.D., Sedjo, R.A. and Reid, J.W. (1994) Valuing Biodiversity for Use in Pharmaceutical Research. Resources for the Future, Washington, DC, May. (also *Journal of Political Economy* (1996) 104(Feb.): 163–185.)

Sing Nijar, G. and Yoke Ling, C. (1994) The implications of the intellectual property rights regime of the convention on biological diversity and GATT on biodiversity conservation: A Third World perspective. In: Krattiger, A.F. *et al.* (eds) *Widening Perspectives on Biodiversity.* IAE and IUCN, Geneva and Gland, Chap. 5.4.

Sittenfeld, A. and Gamez, R. (1993) Biodiversity prospecting by INBio. In: Reid, W.V. *et al.* (eds) *Biodiversity Prospecting: Using Genetic Resources for Sustainable Development.* World Resources Institute, Washington, DC, Chap. III.

Smale, M. (1996) Indicators of Genetic Diversity in Bread Wheats: Selected Evidence on Cultivars Grown in Developing Countries. Paper presented at CEIS-Tor Vergata Symposium on the Economics of Valuation and Conservation of Genetic Resources for Agriculture, Rome, May 13–15.

SSRC (Social Science Research Council) (1996) Meeting the Challenge: Strategic Considerations for Developing Modus Operandi for the Convention on Biological Diversity. New York. August.

Stephenson, D.J. Jr. (1994) A legal paradigm for protecting traditional knowledge. In: Greaves, T. (ed.) *Indigenous Property Rights for Indigenous Peoples.* Society for Applied Anthropology, Oklahoma City, Oklahoma, Chap. 12, pp. 181–189.

Straus, J. (1988) Biotechnology and Its International Legal and Economic Implications. Talk presented at the UN Conference on Trade and Development, Geneva.

Straus, J. (1994) Patenting of human genes and living organisms: the legal situation in Europe. In: Vogel, F. and Grunwald, R. (eds) *Patenting of Human Genes and Living Organisms.* Springer-Verlag, Berlin and New York, pp. 12–29.

Straus, J. and Moufang, R. (1990) *Deposit and Release of Biological Materials for the Purpose of Patent Procedure.* Momos Verlagsgesellschaft, Baden-Baden.

Swaminathan, M.S. and Hoon, V. (1994) *Methodologies for Recognizing the Role of Informed Innovator in the Conservation and Utilization of Plant Genetic Resources.* CRSARD, Proceedings No. 9, Medras.

Swanson, T. (1995) The appropriation of evolution's values: an institutional analysis of intellectual property regimes and biodiversity conservation. In: Swanson, T. (ed.) *Intellectual Property Rights and Biodiversity Conservation.* Cambridge University Press, Cambridge, Chap. 7, 41–75.

Thompson, D.B. (1992) Concepts of Property and the Biotechnology Debate. In: *Ethics and Patenting of Transgenic Organisms.* Occasional Papers No. 1, National Agricultural Biotechnology Council, Ithaca, New York, Sept.

UN (United Nations) (1974) Resolution 3281(XXIX). Sixth Special Session of the UN General Assembly, New York.

UNCTAD (United Nations Conference on Trade and Development) (1975) The Role of the Patent System in the Transfer of Technology to Developing Countries. TD/B/AC-11/19/Rev. 1, New York.

UNCTAD (United Nations Conference on Trade and Development) (1990) *Transfer and Development of Technology in Developing Countries: A Compendium of Policy Issues.* UNCTAD, Geneva.

UNDP (United Nations Development Programme) (1994) Conserving Indigenous Knowledge: Integrating two systems of innovation. New York, 1 Sept.

UNDP (United Nations Development Programme) (1995) Draft guidelines for support to indigenous peoples. Draft project proposal, Nairobi, 31 Aug.

UNEP (1994) Issues on the Conservation of Biological Diversity in Africa. African Ministerial Conference on the Environment, Nairobi, 6 May.

UNEP/CBD/COP/1/4 (1994) Report of the Intergovernmental Committee on the Convention on Biological Diversity, 21 Sept.

UNEP/CBD/COP/2/13 (1995) Access to Genetic Resources and Benefit-Sharing: Legislation, Administrative and Policy Information. Report by the Secretariat, 6 Oct.

UNEP/CBD/COP/2/17 (1995) Intellectual Property Rights and Transfer of Technologies Which Make Use of Genetic Resources. Note by the Secretariat, 6 Oct.

UNEP/CBD/COP/3/20 (1996) Access to Genetic Resources. Note by the Executive Secretary, 5 Oct.

UNEP/CBD/COP/3/22 (1996) The Impacts of Intellectual Property Rights Systems on the Conservation and Sustainable Use of Biological Diversity and on the Equitable Sharing of Benefits from Its Use. Note by the Executive Secretariat, 22 Sept.

UNEP/CBD/COP/3/23 (1996) The Convention on Biological Diversity and the Agreement on Trade-Related Intellectual Property Rights (TRIPs): Relationships and Synergies, 5 Oct.

UNEP (United Nations Environment Programme) (1992) Convention on Biological Diversity. Knowledge, Innovations and Practices of Indigenous and Local Communities. Nairobi.

UNEP, Environmental Law and Institutions Programme Activity Centre (1992) Convention on Biological Diversity. No. 92-8314, June.

UNEP/CBD/IC/2/11 (1994) Report of the Open-Ended Intergovernmental Meeting of Scientific Experts on Biological Diversity, 26 April.

UNEP/CBD/IC/2/13 (1994) Ownership of, and access to, *ex situ* genetic resources: Farmers' rights and rights of similar groups. Note by the Interim Secretariat.

UNEP/CBD/IC/2/14 (1994) Farmers' Rights and Rights of Similar Groups. Note by the Interim Secretariat, 20 May.

UNEP/CBD/SBSTTA/2/Inf.2 (1996) Submissions Received by the Secretariat Concerning the Transfer and Development of Technologies, Sept.

UNEP/CBD/SBATTA/2/6 (1996) Ways and Means to Promote and Facilitate Access to, and Transfer and Development of Technology, Including Biotechnology, 12 Aug.

UNEP/CBD/SBSTTA/2/7 (1996) Knowledge, Innovations and Practices of Indigenous and Local Communities. Note by the Secretariat, 10 August.

UNEP/CBD/SBSTTA/2/13 (1996) Economic Valuation of Biological Diversity. Note by the Secretariat, Montreal, 9 July.

UNOCIAE-C (1994) *Plantas Medicinales del Campo.* Imbabura, Bolivia.

UPOV (International Union for the Protection of New Varieties of Plants) (1992) *Essentially Derived Varieties.* IOM/6/2, Geneva, 17 Aug.

US Department of Commerce, Patent and Trademark Office (1983) *General Information Concerning Patents.* Washington, DC, Feb.

US National Research Council (1992) *Conserving Biodiversity – A Research Agenda for Development Agencies.* National Academy Press, Washington, DC.

van Wijk, J. (1995) Plant breeders' rights create winners and losers. *Biotechnology and Development Monitor* 23(June), 15–19.

van der Walt, W.J. (1994) Brief Review of Intellectual Property Rights in South Africa. Mimeo, South African National Seed Organization, Pretoria.

Vaughan, O., Malanoski, M., West, D. and Handy, C. (1994) *Firm Strategies for Accessing Foreign Markets and the Role of Government Policy.* Working Paper 5/94, Agriculture Canada, Policy Branch, Ottawa, Dec.

Veatch, R. (1976) Three theories of informed consent: philosophical foundations and policy implications. In: US National Commission for the Protection of Human Subjects of Biomedical and Behavioral Research, *Belmont Report*, Department of Health, Education, and Welfare, Pub. No. (OS)78-0014, Washington, DC, Chap. 26.

Vellvé, R. (1989) An overview of concerns regarding the impacts of patenting life forms in the Third World. In: Lesser, W., Straus, J., Duffey, W. and Vellvé, R. (eds) *Equitable Patent Protection for the Developing World.* Cornell University, Staff Paper 89-36, Department of Agricultural Economics, Cornell University, Nov.

Viscusi, W.K., Vernon, J.M. and Harrington, J.E. Jr. (1992) *Economics of Regulation and Antitrust.* D.C. Heath and Co, Lexington, Massachusetts.

Weiss, C., Jr. (1995) *A Proposed New Fund to Promote Value-Added through Bioprospecting.* Working Paper No. 23, Biodiversity/Biotechnology Programme, International Academy of the Environment, Geneva.

Wetter, J.G. and Priem, C. (1991) Costs and Their Allocation in International Commercial Arbitration, *American Review of International Arbitration* 2, 249–349.

Wilkes, G. (1993) *In Situ* Conservation in Guatemala of Teosinte: The Closest Relative of Maize. Unpublished.

Wilkes, H.G. (1988) Plant Genetic Resources over Ten Thousand Years: From a Handful of Seed to the Crop-Specific Mega-Gene Banks. In: Kloppenburg, J.R. Jr (ed.) *Seeds and Sovereignty: The Use and Control of Plant Genetic Resources.* Duke University Press, Durham, North Carolina, Chap. 3.

Wilson, E.O. (1992) *The Diversity of Life.* Belknap, Cambridge.

WIPO (World Intellectual Property Organisation) (1985) *Model Provisions for National Laws on the Protection of Expressions of Folklore against Illicit Exploitation and other Prejudicial Actions.* WIPO, Geneva.

WIPO (World Intellectual Property Organisation) (1990) *Exclusions from Patent Protection*. WIPO, Geneva, HL/CM/INF/1 Rev., May.

WIPO (World Intellectual Property Organisation) (1995) *WIPO Mediation and Arbitration Rules*. Pub. No. 446(E), WIPO, Geneva.

WIPO (World Intellectual Property Organisation) (1996) WIPO Copyright Treaty, CRNR/DC/89, 20 Dec.

World Bank (1994) *World Development Report 1994*. Oxford University Press, Oxford.

Yamin, F. (1995) The Biodiversity Convention and Intellectual Property Rights World Wide Fund for Nature, Gland, Oct.

Young, S. (1989) Testimony on Bill C-15, An Act Respecting PBR. Issue No. 5, House of Commons, Nov. 2.

Yuthavong, Y. and Gibbons, G.C. (1994) *Biotechnology for Development: Principles and Practice Relevant to Developing Countries*. National Science and Technology Development Agency, Bangkok.

Index